Searching for ORDER IN the COMPLEXITY of Evolving Worlds

THE SANTA FE INSTITUTE is the world headquarters for complexity science, operated as an independent, nonprofit research and education center located in Santa Fe, New Mexico. Our researchers endeavor to understand and unify the underlying, shared patterns in complex physical, biological, social, cultural, technological, and even possible astrobiological worlds. Our global research network of scholars spans borders, departments, and disciplines, bringing together curious minds steeped in rigorous logical, mathematical, and computational reasoning. As we reveal the unseen mechanisms and processes that shape these evolving worlds, we seek to use this understanding to promote the well-being of humankind and of life on earth.

[THE EMERGENCE OF PREMODERN STATES]

New Perspectives on the Development of Complex Societies

JEREMY A. SABLOFF
& PAULA L.W. SABLOFF

editors

⁵ᶠᵢ PR⚘·SS

THE SANTA FE INSTITUTE PRESS

1399 Hyde Park Road
Santa Fe, New Mexico 87501

The Emergence of Premodern States: New Perspectives
on the Development of Complex Societies
ISBN (HARDCOVER): 978-1-947864-12-2
Library of Congress Control Number: 2018933822

The SFI Press is supported by the
Feldstein Program on History, Regulation, & Law, the
Miller Omega Fund, and Alana Levinson-LaBrosse.

This publication was made possible through the support
of a grant from the John Templeton Foundation. The opinions
expressed in this publication are those of the authors and do not
necessarily reflect the views of the John Templeton Foundation.

ALSO IN THE SFI PRESS SEMINAR SERIES

History, Big History, & Metahistory
David C. Krakauer, John Lewis Gaddis, & Kenneth Pomeranz, eds.

THE SERIES OF EVENTS that led to the rise of chiefdoms, and eventually to their further development into states, occurred independently in many regions. The fact that this development began only after agriculture had appeared, but that in a short time thereafter essentially the same result was produced in all five continents, suggests that a process, at once simple, yet almost inexorable, had suddenly been triggered. It seems safe to conclude, therefore, that a common set of circumstances, identifiable and recurring, were at work here, yielding similar results along a broad front. What were they?

ROBERT CARNEIRO
The Muse of History and the Science of Culture (2000)

CONTRIBUTORS

Lily Blair, *Stanford University*

R. Kyle Bocinsky, *Washington State University*

Stefani A. Crabtree, *Pennsylvania State University*

Skyler Cragg, *formerly Santa Fe Institute*

Laura Fortunato, *University of Oxford and Santa Fe Institute*

Paul L. Hooper, *Santa Fe Institute*

Hilliard S. Kaplan, *Chapman University*

Timothy A. Kohler, *Washington State University and Santa Fe Institute*

Scott G. Ortman, *University of Colorado Boulder and Santa Fe Institute*

Peter N. Peregrine, *Lawrence University and Santa Fe Institute*

Jeremy A. Sabloff, *Santa Fe Institute and University of Pennsylvania*

Paula L.W. Sabloff, *Santa Fe Institute*

Eric Alden Smith, *University of Washington*

Henry T. Wright, *University of Michigan and Santa Fe Institute*

TABLE OF CONTENTS

Part I: Background

Part II: New Research

Part III: Syntheses

PART I
Background

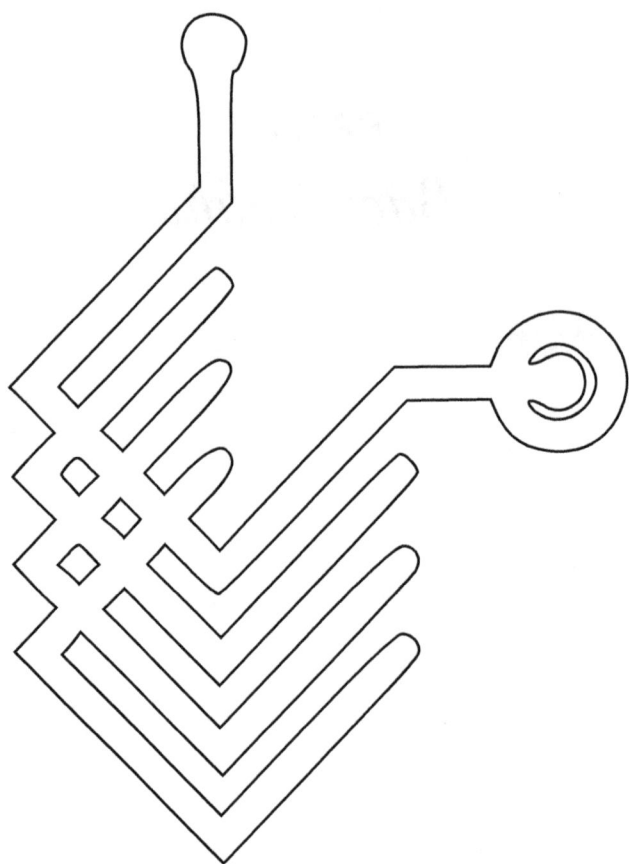

℧

EXTENDING OUR KNOWLEDGE
OF PREMODERN STATES

Jeremy A. Sabloff, Santa Fe Institute and University of Pennsylvania

This volume examines the emergence of premodern states around the globe. In at least Old Kingdom Egypt, Lower Mesopotamia, the Indus Valley, Shang China, Protohistoric Hawai'i, Mesoamerica, and the Andes, societies evolved complex structures without a pre-existing model to follow. Chapter authors undertake comparative analysis to better understand the similarities in processes found in the Old and New Worlds and to build stronger hypotheses about general developments in sociocultural evolution. They also consider secondary states and negative examples—that is, where states did not develop even in seemingly favorable ecological settings. While theoretical interest in the emergence of early states has had a long history in archaeological research, as the authors in this book show, the availability of recent research findings across the world and complexity science methodologies now allow new and more refined thinking than was previously possible on this key topic. Now we can ask more detailed questions about the processes underlying early state formation. What is the appropriate unit of analysis for studying state emergence? What are the implications of changing statuses and roles for elites and followers at the time of state formation? What are the key factors in the evolution of social, political, economic, and religious complexity?

Defining the Early State

In order to address such questions of state formation, which so captivate archaeologists' imaginations, we need to start with a more basic one: How do we know an early state when we see one? That is, what basic characteristics do early states share, and did states undergo similar processual changes to become states? In Part I, we start with Henry Wright's review of archaeologists' working definitions of early states from the nineteenth century to today in order to contextualize the efforts of archaeologists who are grappling with the issues surrounding the emergence of states. Wright discusses the strengths and weaknesses of various approaches, most of which focus on just one or a few processes that mark the transition from agricultural villages to centrally organized states. He suggests that complexity science methods are already greatly refining *how* archaeologists determine the nonlinear interdependence of processes that resulted in the early states. Laura Fortunato (in chapter 3) follows Wright's discussion with a very helpful historical review of comparative interdisciplinary approaches to studying the development of complex societies. In particular, she discusses the utility, as well as some problems with the definitional approaches, of the *Outline of Archaeological Traditions* (Peregrine 2001) and the *Encyclopedia of Prehistory* (Peregrine and Ember 2001–2002), two of the key comparative tools used by archaeologists. If archaeologists are to reach the comparative goals of the authors in this book, then answers to the questions raised in chapter 3 will have to be further refined.

In Part II, Paula Sabloff and Skyler Cragg (chapter 4), examine how status-and-role, a well-established concept in anthropology, can provide a new way to understand archaic states. The authors have adapted anthropologist Ralph Linton's original concept in a comparative analysis of primary and secondary premodern states and have contrasted them with a few non-state traditions in order to answer the following questions: Are state-level societies organized according to the same statuses? Are non-states organized differently? If so, do non-states share the same statuses as states? Do non-state-level societies have people playing the same roles (i.e., having the same rights, responsibilities, and behaviors) as people in state-level societies? What do the patterns say about archaic state organization?

In addressing these questions, the authors offer a new perspective on the shared characteristics of early states by using network analysis to compare the co-occurrence of statuses (positions) with expected roles and observed behaviors. From rulers and gods to slaves and prisoners of war, states exhibit certain patterns that are not shared with non-states. Findings include the role of high-status women in long-term alliance building and warfare. The analysis clearly indicates that Hawai'i was a primary state, ~3~ as Hommon (2013) and Kirch (2010) have long argued. But the early state in Hawai'i did not have cities. So a strong inference can be made that early state growth is not necessarily linked to urban growth, and even more broadly, as Jennings and Earle (2016) have argued, that the processes of urban and state development are different.

New Perspectives on the Processes in State Formation

While scholars now know some of the key processes that resulted in state formation, discussed below, chapters in this book also note issues that arise when one views early state societies as complex systems. They attempt to integrate these processes and relate them to larger patterns. In chapter 7, Scott Ortman, Lily Blair, and Peter Peregrine uncover macrocultural processes by documenting basic patterns in human social evolution in their expanded version of *The Atlas of Cultural Evolution*. They find that all the traditions with state-level organization are best pictured as patterned "networks of matter, energy, and information." These patterns illustrate several basic properties of the evolutionary process and indicate how and why a complex systems perspective is valuable for further progress.

Innovation and Inevitability

Throughout this book, authors seek to determine whether the emergence of complex societies is inherent in human society or reflects instances where humans have created these economies through various forms of innovation. Does the emergence of social complexity involve the realization of a latent potential, as suggested by scaling laws in contemporary cities (see Ortman et al. 2014), or the creation of new potential, as suggested by studies of evolutionary innovation? This is a difficult question to answer for any complex system, but recent perspectives from other sciences suggest that it is the *interaction* of both processes plus favorable conditions that results in the emergence of complexity (e.g., West et al. 1997; Erwin 2008; Arthur 2009). Such new understandings also should allow scholars to examine the extent to which evolutionary innovations and scaling laws contributed to the emergence of early urban states (also see Shennan [2009], among others, for a general evolutionary perspective in archaeology, and O'Brien and Shennan [2009] on innovation and evolutionary thinking).

~4~

When one considers the nature of early state societies from a complex systems perspective, additional issues come into play. One is the interplay between micro and macro processes in the evolution of social complexity. Here, a key question is how the behavior of individuals and households at the microevolutionary level affects long-term patterns that emerge at the macroevolutionary, societal level. Did different forms of social organization in earlier and smaller-scale societies, such as different forms of family structure (e.g., Fortunato 2011), significantly influence their trajectories on the path to increasing scale and complexity? Using agent-based modeling (see Kohler and Gumerman 2000; Kohler and van der Leeuw 2007), a key method of complex systems science, the exciting goal of chapters 5 and 6 by Hooper et al. and Kohler et al., respectively, is to attempt to answer these questions through simulation by discovering the fundamental relationships among key micro and macro variables that shed light on the pathways to the rise of early states. These simulations also speak to the ecological conditions that affect the inevitability of innovation toward states.

Measuring Key Processes in Urban State Formation

Looking more broadly at early states, it is important to briefly situate the chapters in this volume in the intellectual context of current archaeological thinking. In particular, comparative studies have identified five factors that are characteristic of early urban state emergence:

1. Population/resource balances
2. The formation of territorial boundaries within which political elites achieved a monopoly on the use of force
3. A dramatic expansion of specialized production and exchange
4. Significant investments in infrastructure and monuments in political capitals
5. Sanctification of political authority through ideologies that present elites as beings with the power to impact socio-natural forces

Many of the chapters touch on aspects of these factors. While archaeologists first suggested the processes as broad trends, we now know what measurements are necessary to determine their scope, as many of the chapters clearly show.

Population/resource balances: Archaeological research indicates that the rise of early urban states is universally associated with population growth, but that population growth does not always lead to the rise of social complexity. Carneiro (1970; also see Spencer 1998) suggests that a process labeled "circumscription" or "encagement" is a prerequisite for the rise of complex urban states. Strategically,

~5~

this process is concerned with the defensibility of resources, particularly land (Hooper et al. and Kohler et al., chapters 5 and 6 of this volume). In such situations, individuals do not have the option of moving elsewhere to obtain critical resources. To investigate the role of these variables, it is necessary to collect data on the absolute population size of a settlement system, the total area of agricultural production within it, and the basic productivity of food staples grown or collected to provide the bulk of subsistence. There have been significant advances in measurement in this area (see, for example, Kowalewski 2003; Hill et al. 2014; and Kohler et al., chapter 6 in this volume).

Territory formation: In several cases, the emergence of urban states involved the unification and pacification of smaller, previously antagonistic polities and the abandonment of multiple competing centers in favor of a single capital city (Marcus and Flannery 1996; Ortman, chapter 8 in this volume). This was likely achieved through military conquest and maintained by policing. It appears to be the case in non-urban states as well (Kirch 2010). To investigate the role of these variables, it is necessary to devise measures of political competition and warfare. These can be obtained through quantitative summaries of settlement size distributions, the total areas of political systems as defined by settlement and material culture distributions, and frequencies of premortem and perimortem trauma on skeletal remains.

Specialized production and exchange: Urban state formation is often associated with a dramatic increase in economic specialization and the formalization of exchange networks. In some cases, specialized production took place under direct control of political authorities, whereas in others it was independent (Brumfiel and Earle 1987). The extent to which these economic processes encouraged or resulted from urban state formation can be examined by quantifying workshop and market space within settlements;

~6~

estimating proportions of material goods that derive from specific workshops; characterizing the modes of transport for goods and distances over which goods were imported to early cities; and measuring the agricultural hinterlands surrounding such cities (Ortman et al. 2014). Technological innovation rates can also be estimated from the appearance of new artifact types appearing in assemblages.

Investment in public works: Previous studies suggest that investments in public works—temples, administrative buildings, marketplaces, aqueducts, roads, and so forth—were emphasized during periods of initial state emergence (e.g., Mendelssohn 1974). To investigate the role of such investments, it is necessary to quantify their energetic costs for material acquisition and construction (e.g., Abrams 1994; Abrams and Bolland 1999).

~7~

Sanctification of authority: Most cases of early state formation are associated with iconographic and architectural evidence suggesting new forms of persuasion that cemented the political authority of rulers (e.g., Marcus and Flannery 1996). This basic pattern suggests that, in addition to political, military, and economic innovations, primary state formation involved innovations in ideology as well. The ideological dimension of urban state emergence can be examined through the identification of specific analogies related to political authority in the material culture, oral traditions, and written records (Ortman 2000). Although they may only be quantifiable on a presence/absence basis, such data collected across a range of ancient societies should be sufficient to identify patterns in the analogies used in premodern state ideologies vs. societies that did not make the transition to statehood (see Sabloff and Cragg, chapter 4 in this volume, as one example of the productive use of presence/absence data).

Conclusion

Empirical data about the key evolutionary transition from agricultural village to urban-centered state have proliferated over the past few years through a host of research projects. In this respect, archaeology is undergoing a change comparable to the data deluge in many of the other sciences. Despite much attention and interest, however, the construction and testing of theoretically driven explanations for the emergence of the urban state have lagged behind.

Until recently, one of the reasons for this lag has been an emphasis on the uniqueness of early states in different parts of the world and the paucity of good, detailed, theoretically based comparative studies, as archaeologists focused on site-specific excavation and regional settlement survey (early exceptions include Childe [1951], Adams [1966], and Renfrew [1972]). However, in recent years a growing number of useful comparative studies have appeared (e.g., Feinman and Marcus 1998; Johnson and Earle 2000; Trigger 2003; Spencer and Redmond 2004; Marcus and Sabloff 2005; Blanton and Fargher 2008; Maisels 2010; Flannery and Marcus 2012; also see Smith 2012). These have identified a number of causal variables in the emergence of complex urban states. The chapters in this book build upon these important insights.

Other factors behind this lack of theoretical progress in recent years include the slowness of publication of the results of archaeological projects on premodern states and the difficulty of access to unpublished data, thus inhibiting synthetic attempts (see Altschul et al. 2018). In addition, the obvious complexity of the processes involved in the rise of states hinders major theory building.

Perhaps most important, theory building has lagged behind other disciplines because many of the comparative methodologies used thus far lack the sophistication to determine the emergence of social complexity. But this trend has turned a corner, and many archaeologists today have adopted a complex systems perspective—pioneered at the Santa Fe Institute—that involves fully interdisciplinary approaches to the study of individual complex societies, including landscape studies, ecological studies, and a host of chemical and biological studies. In addition, many archaeologists are using quantitative modeling to simulate and clarify the relevant evolutionary processes (see, e.g., Spencer 1998; Hooper et al. 2010; Gavrilets et al. 2010; Kohler et al. 2012; Wilkinson et al. 2013). Given the recent accumulation of accessible data on

~9~

Many archaeologists today have adopted a complex systems perspective that involves fully interdisciplinary approaches to the study of individual complex societies, including landscape studies, ecological studies, and a host of chemical and biological studies

early urban states (see Fortunato, chapter 3, for a critical overview of the main sources and summaries of these data), coupled with the reemergence of comparative studies, we are now in a position to make significant theoretical advances concerning this key episode of human social organization that provided the foundations of the contemporary world. For example, in Part III, chapter 8, Scott Ortman, inspired by breakthroughs in evolutionary biology and psychology, develops a novel model of social evolution via the spread of cultural metaphors (cultural genotypes) that promote social coordination and cohesion. He suggests that cultural genotypes define and redefine social distance between kin and nonkin, thereby making nonkin cohesion possible. Ortman notes that while human societies are complex networks of people, energy, and information, our understanding of how humans actually represent information remains poorly developed. To begin to rectify this problem, he builds a model of information representation that is grounded in experimental research in cognitive and social psychology. The model exhibits the key properties of genotypes noted in studies of evolutionary innovation. This framework also leads to several insights concerning the evolution of social complexity. Finally, in chapter 9, Peter Peregrine offers an important argument that evolutionary biology can help illuminate the reasons for recurrent social formations—and, in particular, the premodern state. He also looks at the issue of cultural taxonomy and argues that traditional anthropological taxonomies of cultural evolution, including the concept of the state, are both useful and appropriate.

In sum, the complexity science approach to early states, as seen in the chapters of this volume, will, we hope, move us closer toward our goal: the understanding of how and why early states emerged independently in so many parts of the world and did not emerge in others. ✸

REFERENCES CITED

Abrams, Elliot M.
   1994   *How the Maya Built Their World: Energetics and Ancient Architecture.* University of Texas Press, Austin.

Abrams, Elliot M., and Thomas Bolland
   1999   Architectural Energetics, Ancient Monuments, and Operations Management. *Journal of Archaeological Method and Theory* 6:263–290.

Adams, Robert McC.
   1966   *The Evolution of Urban Society.* Aldine, Chicago.

Altschul, Jeffrey H., Keith W. Kintigh, Terry H. Klein, William H. Doelle, Kelley A. Hays-Gilpin, Sarah A. Herr, Timothy A. Kohler, Barbara J. Mills, Lindsay M. Montgomery, Margaret C. Nelson, Scott G. Ortman, John N. Parker, Matthew A. Peeples, and Jeremy A. Sabloff
   2018   Fostering Collaborative Synthetic Research in Archaeology. *Advances in Archaeological Practice* 6(1):19–29.

Arthur, W. Brian
   2009   *The Nature of Technology: What It Is and How It Evolves.* Free Press, New York.

Blanton, Richard, and Lane Fargher
2008   *Collective Action in the Formation of Pre-Modern States.* Springer, New York.

Brumfiel, Elizabeth, and Timothy Earle
1987   *Specialization, Exchange, and Complex Societies.* Cambridge University Press, Cambridge.

Carneiro, Robert L.
1970   A Theory of the Origin of the State. *Science* 169:733–738.

Childe, V. Gordon
1951   *Social Evolution.* Meridian Books, New York.

Erwin, Douglas H.
2008   Macroevolution of Ecosystem Engineering, Niche   ~II~ Construction and Diversity. *Trends in Ecology and Evolution* 23:304–310.

Feinman, Gary M., and Joyce Marcus (editors)
1998   *Archaic States.* School of American Research Press, Santa Fe, New Mexico.

Flannery, Kent V., and Joyce Marcus
2012   *The Creation of Inequality.* Harvard University Press, Cambridge, Massachusetts.

Fortunato, Laura
2011   Reconstructing the History of Marriage and Residence Strategies in Indo-European-Speaking Societies. *Human Biology* 83:129–135.

Gavrilets, Sergey, David G. Anderson, and Peter Turchin
2010   Cycling in the Complexity of Early Societies. *Cliodynamics* 1(1).

Hill, Kim R., Brian M. Wood, Jacopo Baggio, A. Magdalena Hurtado, and Robert T. Boyd
2014   Hunter-Gatherer Inter-Band Interaction Rates: Implications for Cumulative Culture. *PLoS ONE* 9(7):e102806.

Hommon, Robert
2013   *The Ancient Hawaiian State.* Oxford University Press, Oxford.

Hooper, Paul L., Hilliard S. Kaplan, and James L. Boone
2010   A Theory of Leadership in Human Cooperative Groups. *Journal of Theoretical Biology* 265(4):633–646.

Jennings, Justin, and Timothy Earle
2015   Urbanization, State Formation, and Cooperation: A Reappraisal. *Current Anthropology* 57:474–493.

Johnson, Allen W., and Timothy Earle
2000   *The Evolution of Complex Societies.* 2nd ed. Stanford
University Press, Stanford, California.

Kirch, Patrick
2010   *How Chiefs Became Kings: Divine Kingship and the Rise
of Archaic States in Ancient Hawai'i.* University of California Press,
Berkeley.

Kohler, Timothy A., Denton Cockburn, Paul L. Hooper, R. Kyle
Bocinsky, and Ziad Kobti
2012   The Coevolution of Group Size and Leadership: An Agent-
Based Public Model for Prehispanic Pueblo Societies. *Advances in
Complex Systems* 15.

Kohler, Timothy A., and George Gumerman (editors)
2000   *Dynamics in Human and Primate Societies: Agent-Based
Modeling of Social and Spatial Processes.* Santa Fe Institute and
Oxford University Press, Santa Fe, New Mexico.

Kohler, Timothy A., and Sander van der Leeuw (editors)
2007   *Model-Based Archaeology of Socionatural Systems.* SAR
Press, Santa Fe, New Mexico.

Kowalewski, Stephen
2003   The Evolution of Complexity in the Valley of Oaxaca.
Annual Review of Anthropology 19:39–58.

Maisels, Charles
2010   *The Archaeology of Politics and Power: Where, When, and
Why the First States Formed.* Oxbow Books, Oxford.

Marcus, Joyce, and Kent V. Flannery
1996   *Zapotec Civilization: How Urban Society Evolved in Mexico's
Oaxaca Valley.* Thames and Hudson, London.

Marcus, Joyce, and Jeremy A. Sabloff (editors)
2005   *The Ancient City: New Perspectives on Urbanism in the Old
and New World.* SAR Press, Santa Fe, New Mexico.

Mendelsohn, Kurt
1974   *The Riddle of the Pyramids.* Praeger, New York.

O'Brien, Michael J., and Stephen J. Shennan (editors)
2009   *Innovation in Cultural Systems: Contributions from
Evolutionary Anthropology.* MIT Press, Cambridge, Massachusetts.

Ortman, Scott G.
2000   Conceptual Metaphor in the Archaeological Record:
Methods and an Example from the American Southwest. *American
Antiquity* 65(4):613–645.

~12~

Ortman, Scott G., Andrew H. F. Cabaniss, Jennie O. Sturm, and Luis M. A. Bettencourt
2014   The Pre-History of Urban Scaling. *PLoS* ONE 9(2):e87902. DOI:87910.81371/journal.pone.0087902.

Peregrine, Peter N.
2001   *Outline of Cultural Traditions.* HRAF Press, New Haven, Connecticut.

Peregrine, Peter N., and Melvin Ember (editors)
2001–2002   *Encyclopedia of Prehistory.* Kluwer Academic/Plenum Publishers, New York.

Renfrew, Colin
1972   *The Emergence of Civilization: The Cyclades and the Aegean in the Third Millennium B.C.* Methuen, London.

Shennan, Stephen J.
2009   *Pattern and Process in Cultural Evolution.* University of California Press, Berkeley.

Smith, Michael E. (editor)
2012   *The Comparative Archaeology of Complex Societies.* Cambridge University Press, Cambridge.

Spencer, Charles S.
1998   A Mathematical Model of Primary State Formation. *Cultural Dynamics* 10:5–20.

Spencer, Charles S., and Elsa M. Redmond
2004   Primary State Formation in Mesoamerica. *Annual Review of Anthropology* 33:173–199.

Trigger, Bruce G.
2003   *Understanding Early Civilizations: A Comparative Study.* Cambridge University Press, Cambridge.

West, Geoffrey B., Brian J. Enquist, and James H. Brown
1997   A General Model for the Origin of Allometric Scaling Laws in Biology. *Science* 276:122.

Wilkinson, T. J., M. Gibson, and M. Widell (editors)
2013   *Models of Mesopotamian Landscapes: How Small-Scale Processes Contributed to the Growth of Early Civilizations.* Archaeopress, Oxford.

~13~

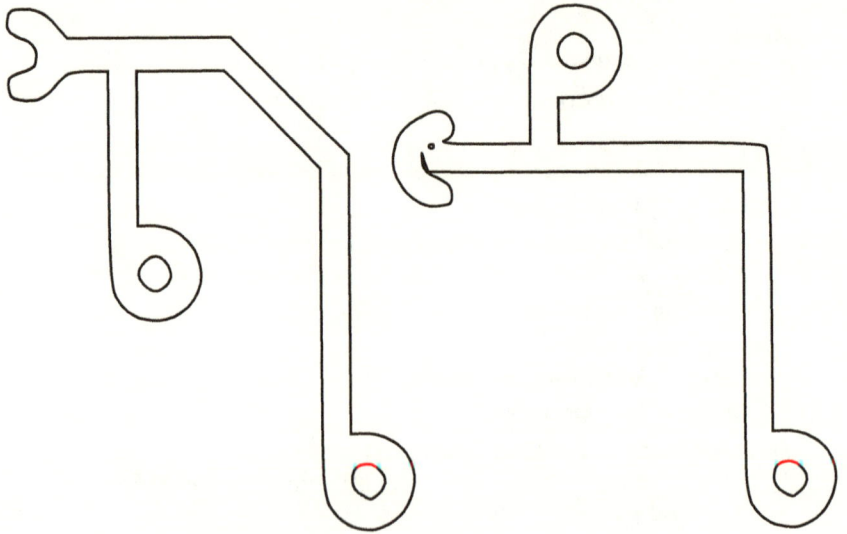

6

THE PROBLEM OF STATES:
THE STATE OF THE PROBLEM

Henry T. Wright, University of Michigan and Santa Fe Institute

Everyone who reads this chapter lives under the control of a state. We benefit from the security the state provides, from specialized organizations that deliver education at the beginning of life and health care at its end. We may be forced to pay taxes to sustain means of suppression we do not like, and we may be forced to fight in wars we do not accept. It is not surprising that scholars have been fascinated with the rise and spread of these polities during the last six millennia.

Early Studies

The first theories of state formation are constructed explanations of specific trajectories often found in epic poems or in mythohistoric texts. They focus on the founder's character, whether it is the physically perfect and semidivine Gilgamesh, the ruler of Uruk in early third millennium BCE Sumer (George 1999), or the diligent and very human Yu of the early second millennium BCE Xia dynasty in central China (Nienhauser 1994). However, these actors do deal with the fundamental dynamic processes important in their worlds. The accounts of Yu focus on his efforts to organize farmers to build dikes and canals to control floods; the earlier portions of *The Epic of Gilgamesh*, which focus on the wars between Uruk and the city of Kish, seem to detail negotiations between councils of elders and councils of young warriors. If such sources can be taken as early efforts to understand state emergence, it must not be forgotten that these are literary compositions. Explicit efforts to verify accounts and frame explanatory constructs about early state formation come

~15~

later, with the first historians of classical eras such as Thucydides in Greece and Sima Qian in China. Constructs based on comparative studies of different cases considering a diversity of interacting variables do not appear before the thirteenth century CE with the *Muqaddimah of Ibn Khaldun* (1958).

However, the early studies (and indeed all those up to the late nineteenth century) suffer from common weaknesses rooted in the nature of their sources. All used both written documents and oral—often eyewitness—testimony. Documents are written by someone with a point of view and are conserved by others, perhaps

---

Historians are skilled at accounting for text biases and constructing a plausible narrative from the available evidence, but these studies are not data on the operation of whole systems, and they can only be used to evaluate ideas about the emergence of new forms of organization with many caveats.

---

with a different point of view; oral accounts, especially eyewitness accounts, also have a point of view, and suffer further from the fact that a witness can only be at one place at one time and may have a limited understanding of complicated events. Historians are skilled at accounting for text biases and constructing a plausible narrative from the available evidence, but these studies are not data on the operation of whole systems, and they can only be used to evaluate ideas about the emergence of new forms of organization with many caveats.

Twentieth-Century Studies

Scholars in the twentieth century who were interested in the first appearance of complex political systems faced new problems and new opportunities. Relatively early in the century, archaeologists began to realize that complex sociopolitical formations first emerged long before writing systems had developed sufficiently to the point where they could record events, decisions, and the motivations of ancient actors. Gilgamesh, for example, lived a millennium after the first appearance of such formations at the beginning of the Uruk period in the early fourth millennium BCE, and the written accounts we have of him are primarily texts from a millennium later. The problem of the lack of contemporary written documents, however, was to some extent balanced by the results of new approaches to prehistoric and protohistoric archaeology. In all the areas where the earliest civilizations had left durable remains, archaeologists had worked out successive periods and characterized the architecture and artifacts of their major centers. The garbage from which the archaeological record is constructed does not have a point of view, and one might think that, whatever the difficulties of inferring economic, social, and political information from broken things, archaeology should provide a solid understanding of past processes unbiased by the points of view of those who wrote and archived accounts in later times.

~17~

Alas, the early- and mid-twentieth-century archaeologists, historians, and anthropologists who used the newly available but still crude archaeological record faced other problems. In the first place, although garbage had no point of view, the archaeologists certainly did. Most were trained in elite institutions in Europe and the Americas and were employed by great museums and universities that competed for the most striking display materials. It is not surprising that they searched for and excavated major centers and then focused on the temples, palaces, and cemeteries in and around these centers. Given this weakness, it is a tribute to their broad knowledge of history and ethnology that leading synthesists could transcend the problem. In a worldwide comparative study of early civilizations, anthropologist Julian Steward (1949) argued that population growth, irrigation agriculture, and warfare were key processes in all cases. It was Steward who first proposed using the new method of regional archaeological survey to better assess population sizes, canal construction, and fortifications, a challenge that was taken up by Gordon Willey (1953) in Peru and Robert McC. Adams in Mesopotamia (1966). Using only Old World cases, viewed in a Marxist perspective, V. Gordon Childe (1952) argued for interaction of such variables as population growth, production specialization, trade in scarce materials, class emergence, religious centralization, increased record keeping, and increasingly orga-nized warfare as widespread processes in many cases. Using only New World cases, Gordon Willey (Willey and Phillips 1958) looked at the full sweep of development, emphasizing similarities between distant regions, but cautiously avoided interpretations of past processes.

In the second place, these early efforts, however astute they were, all suffered from a poor understanding of chronology. Lacking precise methods to establish the dates of archaeological features, they all used periods of 200–500 years, observational units that included many different historical moments varying from rapid growth to stability to conflict and reorganization. (The exception was the North American Southwest, where tree-ring dating accurate to within a year was developed early in the twen-tieth century. Unfortunately, this was an area where economic and political systems remained relatively small in scale.) In most areas, the coarse units of time created impressions of gradual growth. This, for example, led Elman Service (1975) to propose that the emergence of early states was a long, gradual, and peaceful process in which temple institutions were the main focus of innovation. In contrast, later state emergences studied by ethnohistorians and ethnologists—polities that emerged in interaction with the new modern world system—appeared rapidly under conditions of pervasive violence. We know today that most early cases of state emergence were also rapid and violent.

By the second half of the twentieth century, social and cultural anthropologists were making more rigorous ethnological comparisons using more carefully constructed samples based on a richer ethnographic record than was available to their nineteenth-century predecessors. Marshall Sahlins's (1958) study of the radiation of Polynesian-speaking peoples into the Pacific, which attempted to measure the effect of adaptation to islands of differing diversity and richness, was followed by Irving Goldman's (1970) study of the emergence of different types of leadership in the same vast area. Donna Taylor (1975) undertook a similar and important study of the radiation of Bantu-speaking peoples in sub-Saharan Africa. Taylor was the first to code sociopolitical attributes and use scaling and clustering statistics, and these showed that there were several types of sociopolitical systems, rather than a gradual transition from simple to complex. Though not published in full, Taylor's work was widely read and influential. Interestingly, a recent coding and multivariate study of Pacific peoples by Currie et al. (2010) and comparisons of Bantu and Pacific peoples by Walker and Hamilton (2010)—both using larger samples, historical linguistic controls over developmental trajectories, and sophisticated statistics—arrive at similar conclusions. Though the ethnographic record should not be used to reconstruct "conjectural history," these analyses support the argument that there were abrupt and rapid transitions from one sociopolitical form to another.

By the end of the twentieth century, archaeologists had many new tools with which to study the emergence of the earliest

~19~

complex formations during the Holocene. Most important, time could be measured in many ways with greater precision. Tree-ring sequences are now available in many parts of the world, not just a few arid regions—and even if the datable species are rarely found on archaeological sites, tree-ring sequences can be used to monitor year-to-year changes in humidity, solar insolation, or temperature, important in new forms of modeling cultural trajectories (discussed below). Radiocarbon dates can be corrected for various contaminants and calibrated to astronomical time, and accelerator

techniques allow the dating of much smaller items. In addition, the measure of photons trapped in crystals in ceramics, and more recently in soils, now provides a way of directly measuring the time since archaeological ceramics were last fired and since the sun last shined on an archaeological layer. Other more precise chronological methods have been developed recently.

In addition, neutron activation analysis can identify the sources of many kinds of materials with parts-per-billion characterizations of trace elements. Stones, metals, ceramics, shell, and bone have been successfully ascribed to source, which helps archaeologists elucidate procurement and transport over often-vast distances. Less precise but more portable means of trace-element characterization are now available, and completely different types of characterization, such as stable isotopes, are widely used.

New ways to monitor the relations between biological populations are also becoming increasingly feasible in the past few years. Ancient DNA can be extracted not only from human bone and other tissues but also from those of domestic and commensal animals and from dry-preserved plant remains.

Last, the flood of information provided by new methods requires better programs for data storage and access. More portable information-processing devices with large memories that were inconceivable a few years ago and new forms of software not only for data storage but also for modeling, statistical evaluation, and graphic presentation are now widely available.

The Objects of Study

No amount of observation can resolve problems of past action and organization unless the questions asked of data are clearly phrased. Scholars could not proceed to a more comprehensive understanding of the crucial transition from village life to complex societies without rethinking their basic concepts. From the late 1960s to the 1980s, there was a rich ferment of discussion of both definitions of phenomena and possible explanations. Though some scholars have been dismissive of definitions, it is the case that poorly defined phenomena are difficult to explain. Some terms— for example, *civilization*—specify general areas of interest, and no one seriously thinks they can be "explained." However, the emergence or the behavior of phenomena, which scholars do want to explain or account for, requires careful specification.

~21~

For instance, the concepts of *city* and *state* have both been discussed at length. A city is often conceived of as a large population agglomeration. The difficulty with this is where to place the minimum limit. Geographers long ago handled this difficulty by shifting their focus to *urban systems* and looking at the distribution of sizes of all settlements and correlations between settlement size and other variables (Haggett 1966). Whether one looks at all settlements or only at a subset of larger settlements, many variables are strongly interrelated (Bettencourt et al. 2010), and general theories explaining the relations have been proposed (Bettencourt 2013). Urban systems have also been defined in terms of a kind of economy, one in which there is complementary specialization of the production and distribution of life-sustaining goods (Wright 1969), a construct that has no implications about cities. In addition, *urban society* has been defined in terms of class organization (Adams 1966) that likewise has no implications about cities.

A *state* has traditionally been conceived as a polity in which the political leaders control a legitimate monopoly of force. We owe this definition—or at the least the popularity of this definition—to the fact that many early scholars concerned with states were trained as lawyers. Anthropologists and historians seeking to use this definition look at the existence of formal police roles, codes of law, and legal procedures. This definition has several problems. First, there are many kinds of force in a social formation, and leaders can declare them legitimate or illegitimate as they wish. Thus, there is a danger of subjectivity in determining what is or is not a state. Second, because the monopoly of force must be used when ethical rules and norms are not obeyed, one could not tell whether a well-run state was a state or not. It is probably not a good idea to define a political form in terms of a behavior used when it does not operate smoothly. More positive criteria are needed, and indeed they have been widely discussed. Scholars have often spoken of *professional government* or *specialized administration* when defining states. Discussion often focuses, on the one hand, on roles that assist and amplify the control of paramount rulers and, on the other hand, on the complementarity and interdependence of roles such that none can do their job without the help of others.

Seeking to avoid a commitment to the concept of role (simply because roles are not well defined in the earliest stages of state consolidation, even though the systems of interrelated activities are well defined), I have written about a central control apparatus whose activities as a whole are separate from those of other apparatuses or "subsystems," termed "external specialization," and whose internal activities are different from and dependent on each other, termed "internal specialization" (Wright 1977:381–385; also see Wright 1994). Many political systems have external specialization

~22~

of a central control apparatus, but only states have both external and internal specialization.

Defining states in terms of specialized control activities does not mean that they can be easily recognized in the archaeological record. In some cultural traditions, decision-making (at least in the area of the control of goods and information) is directly attested. Mesopotamianists are fortunate to have the direct evidence of seals and sealing, often found in contexts allowing organizational inferences (Frangipane 2007). Iconographic evidence of control activities is more widespread but harder to interpret. Architectural evidence of the loci of control activities is even more widespread but also not easy to interpret. The use of settlement hierarchies to recognize hierarchies of sociopolitical control is always possible but is suggestive and never definitive. The argument that if there are four or more levels of control (three above the level of producing villages), then there must be stable and complementary specialized control activities or otherwise the system would fission, may be correct, but does a settlement hierarchy recognized from archaeological surveys actually house a control hierarchy? One solution to this problem is to study hierarchies in their own right, without regard for whether they are hierarchies of specialized control operations. Gregory Johnson has developed this perspective, proposing a general explanation of the formation of different kinds of hierarchies because of scalar stress (Johnson 1982) and exemplifying this with an archaeological case study from the Uruk period of the fourth millennium BCE in southwestern Iran (Johnson 1973, 1987).

~23~

Explaining the Emergence of States and Urban Systems

Many explanations of the emergence of states and urban centers have been proposed and evaluated with historical, ethnographic, and archaeological evidence. Let us consider a sampling of a few that might be usefully tested with the more comprehensive datasets that the Santa Fe Institute's Rise of the Archaic State Project developed (see, for example, chapters 4 and 7 in this volume).

Those who define states in terms of a monopoly of force have often focused on explanations involving internal conflicts between classes over wealth and status and the creation of states to consolidate the control of some classes over others. Such constructs were elaborated in the nineteenth century but are still influential. These constructs suffer from the lack of evidence for major flows of fungible wealth (as opposed to wealth in goods that signal status and rank), which only arise with established states and urban systems. While perhaps not useful for the earliest states, which arise out of competition between non-state formations that lack wealth-based classes, those flows may be useful for later state formations, which arose on the peripheries of established states that had created such flows of wealth. Indeed, the cases well known to the pioneers of evolutionary thought in the nineteenth century—such as Morgan (1909 [1877]) and Engels (1902 [1884])—from the Levant, Greece, and Italy were developments peripheral to long-standing states, and it is not surprising that the conflict of classes over the flow of wealth would be a central process in these cases.

Those who define states in terms of specialized or professional governments have evaluated a range of explanations. Some have focused on the stresses created by increasingly complex internal activities. The coordination of different aspects of material procurement and food and craft production and distribution are factors that are hard to evaluate without elaborate economic models, which are only now being created (Wilkinson et al. 2007). However, the specific stresses created by irrigation agriculture—both the demand for management in construction and maintenance and opportunities for control of people presented by control of the water supply—have been thoroughly discussed and tested (Wittfogel 1957; Hunt 1988, 2007). While irrigation is important in cases where it is practiced, major expansion of irrigation systems beyond what small social units can create and maintain develops long after states emerge.

Those who have focused on urbanism as a phenomenon of the increasing scale of diverse flows are more concerned with the effects of scale than with how urban systems emerge. Nonetheless, the question can be posed, and scaling theorists are working on the issue.   ~25~

Other explanations that called on conflict between non-state polities, often driven by local population growth and competition (Carneiro 1970), have also been widely tested using population estimates (based on archaeological surveys) and assessments of settlement fortification and destruction (based on excavation). In some ways, evidence of widespread population growth beginning in the early Neolithic, derived from both osteological samples cleverly processed to yield demographic rates (Bocquet-Appel and Bar-Yosef 2008) and more traditional inference based on numbers and sizes of habitation sites, makes such suggestions inherently plausible. While large populations are necessary for states to function, though, case studies indicate that populations fluctuate widely before state formation and that growth can be a result of immigration to join the emerging successful state polity. Case studies argue for the importance of conflict and military logistics and of actual battles that do demand or select for specialized hierarchical organization, but explanations based on such criteria are also hard to evaluate without complex models, which are only beginning to appear (e.g., Gavrilets et al. 2010 and Kohler et al. in chapter 6 in this volume for smaller political formations).

Those who have focused on urbanism as a phenomenon of the increasing scale of diverse flows are more concerned with the effects of scale than with how urban systems emerge. Nonetheless, the question can be posed, and scaling theorists are working on the issue (Bettencourt 2013).

Those who focus on urbanism as a phenomenon of economic organization have a longer history of research. Even Childe (1952) presented an elegant discussion of the relation between agricultural production, craft production, and the rise of a managerial elite, which could have been formally modeled had the computing technology been available. So far, such economic models have been case-specific, but more general formulations are certainly within reach.

To summarize this brief discussion of efforts to explain the emergence of states and urban systems, proposed explanations focusing on one or a few key variables always seem to require more variables. More important, they require new ways of thinking about relationships and processes. The most promising candidate explanations—whether they involve the control of resources and the profits of exchange, collaboration and coordination of mass labor, or the organization and logistical demands of conflict between polities—all require new ways of thinking about decision-making and about the creation of new symbols and symbolic interrelations. How such novel information is acquired and put to work are fundamental challenges for the next generation of theory builders and of researchers who will test these new constructs. ⚡

POSTSCRIPT

This necessarily brief overview serves as a background for the research presented in this volume. The following chapters present efforts to go beyond these existing lines of research, both in the development of databases useful in comparative studies and in the construction of new explanations. Because research will surely continue and advance, these chapters serve as an interim report of ongoing studies.

REFERENCES CITED

Adams, Robert McC.
   1966   *The Evolution of Urban Society*. Aldine, Chicago.
Bettencourt, Luís M. A.
   2013   The Origins of Scaling in Cities. *Science* 340(6139):1438–1441. DOI:10.1126/science.1235823.
Bettencourt, Luís M. A., José Lobo, Deborah Strumsky, and Geoffrey B. West
   2010   Urban Scaling and Its Deviations: Revealing the Structure of Wealth, Innovation and Crime across Cities. *PloS* ONE 5(11):e13541. DOI:10.1371/journal.pone.0013541.
Bocquet-Appel, J.-P., and O. Bar-Yosef (editors)
   2008   *The Neolithic Demographic Transition and Its Consequences*. Springer, Dordrecht, the Netherlands.
Carneiro, Robert L.
   1970   A Theory of the Origin of the State. *Science* 169:733–738.
Childe, Vere Gordon
   1952   The Birth of Civilization. *Past and Present* 2:1–10.
Currie, Thomas E., Simon J. Greenhill, Russell D. Gray, Toshikazu Hasegawa, and Ruth Mace
   2010   Rise and Fall of Political Complexity in Island Southeast Asia and the Pacific. *Nature* 1467:801–804. DOI:10.1038/nature09461.

Engels, Friedrich
1902 [1884]   *The Origin of the Family, Private Property and the State.* Charles Kerr and Co., Chicago.

Frangipane, Marcella (editor)
2007   *Arslantepe—cretulae: An Early Centralised Administrative System Before Writing.* Università di Roma "La Sapienza," Dipartimento di Scienze Storiche Archeologiche e Antropologiche dell'Antichità, Rome.

Gavrilets, Sergey, David G. Anderson, and Peter Turchin
2010   Cycling in the Complexity of Early Societies. *Cliodynamics* 1:58–80.

George, Andrew
1999   *The Epic of Gilgamesh.* Allen Lane, London.

Goldman, Irving
1970   *Ancient Polynesian Society.* University of Chicago Press, Chicago.

Haggett, Peter J.
1966   *Locational Analysis in Human Geography.* Edward Arnold, London.

Hunt, Robert C.
1988   Size and the Structure of Authority in Canal Irrigation Systems. *Journal of Anthropological Research* 44(4):335–355.
2007   Communal Irrigation: A Comparative Perspective. In *A World of Water,* edited by Peter Boomgard, pp. 187–208. Royal Netherlands Institute for Southeast Asian and Caribbean Studies, Leiden.

Ibn Khaldun
1966   *The Muqaddimah: An Introduction to History.* Translated by Franz Rosenthal. Pantheon, New York.

Johnson, Gregory A.
1973   *Local Exchange and Early State Development in Southwestern Iran.* Anthropological Paper, Museum of Anthropology No. 51. University of Michigan Museum of Anthropology, Ann Arbor.
1982   Organizational Structure and Scalar Stress. In *Theory and Explanation in Archaeology,* edited by Colin Renfrew, Michel Rowlands, and Barbara Segraves, pp. 389–420. Academic Press, Orlando, Florida.
1987   The Changing Organization of Uruk Organization on the Susiana Plain. In *The Archaeology of Western Iran,* edited by Frank Hole, pp. 107–139. Smithsonian Institution Press, Washington, DC.

Morgan, Lewis Henry
1909 [1877]   *Ancient Society.* Charles Kerr and Co., Chicago.

Nienhauser, William H. (editor)
1994   *Shiji: The Grand Scribe's Records.* Indiana University Press, Bloomington.

Sahlins, Marshall
1958   *Social Stratification in Polynesia.* University of Washington Press, Seattle.

Service, Elman
1975   *Origins of the State and Civilization: The Process of Cultural Evolution.* Norton, New York.

Steward, Julian   ~29~
1949   Cultural Causality and Law: A Trial Formulation of the Development of Early Civilization. *American Anthropologist* 51:1–27.

Taylor, Donna
1975   Some Locational Aspects of Middle-Range, Hierarchical Societies. PhD dissertation, Department of Anthropology, City University of New York. Proquest.

Walker, Robert S., and Marcus Hamilton
2010   Social Complexity and Linguistic Diversity in the Austronesian and Bantu Population Expansions. *Proceedings of the Royal Society B* (20 October). DOI:10.1098/rspb.2010.1942.

Wilkinson, Tony J., J. H. Christiansen, Jason Ur, Marcus Widell, and Mark Alaweel
2007   Urbanization within a Dynamic Environment: Modeling Bronze Age Communities in Upper Mesopotamia. *American Anthropologist* 109(1):52–68.

Willey, Gordon R.
1953   *Prehistoric Settlement Patterns in the Viru Valley, Peru.* Bureau of American Ethnology, Bulletin 155. Smithsonian Institution, Washington, DC.

Willey, Gordon R., and Phillip Phillips
1958   *Method and Theory in American Archaeology.* University of Chicago Press, Chicago.

Wittfogel, Karl
1957   *Oriental Despotism.* Yale University Press, New Haven, Connecticut.

Wright, Henry T.

1969   *The Administration of Rural Production in an Early Mesopotamian Town.* University of Michigan Museum of Anthropology Anthropological Papers No. 38. Museum of Anthropology, Ann Arbor.

1977   Recent Research on the Origin of the State. *Annual Review of Anthropology* 6:379–397.

1994   Pre-state Political Formations. In *Chiefdoms and Early States in the Near East,* edited by Gil Stein and Mitchell Rothman, pp. 67–84. Revised and reprinted. Madison: Prehistory Press. Originally published 1984, in *The Evolution of Complex Societies: The Harry Hoijer Lectures for 1982,* edited by T. K. Earle. Undena Press, Malibu.

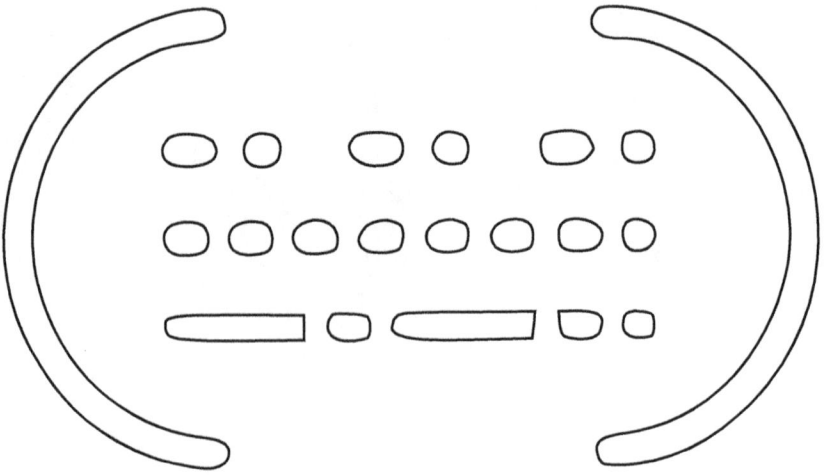

）
╱

SYSTEMATIC COMPARATIVE APPROACHES TO THE ARCHAEOLOGICAL RECORD

Laura Fortunato, University of Oxford and Santa Fe Institute

Increasingly, interdisciplinary research teams are coming together to try to establish regularities, over space and time, in the complex system that is the human phenomenon (see, for example, Kohler et al., chapter 6 in this volume). Although vocabulary and tools ~33~ have changed, the questions that animate this research program bear striking similarity to those pursued by nineteenth-century intellectuals in a quest to establish universal laws shaping human affairs. In fact, that very quest provided the impetus for the emergence of what would later become distinct disciplines in the social and historical sciences, including anthropology[1] and sociology (see Carneiro 2003; Harris 2001; Trigger 2006).

Why, then, is this interdisciplinary research program often met with skepticism, or even outright resistance, within anthropology?

In this chapter I provide a brief outline of developments in the history of anthropology leading to this state of affairs, in the hope of alleviating misunderstanding between those who support the interdisciplinary research program and those who oppose it. As a practical contribution toward this end, I then provide an overview of key established resources for systematic comparative approaches to the archaeological record. I conclude by discussing challenges and opportunities in this area at the interface with recent developments in related archaeological practice.

---

[1] I refer to anthropology as traditionally practiced in North American universities, encompassing archaeology as a subfield.

Historical Sketch

In large part, the current state of affairs in anthropology can be attributed to the prevailing theoretical paradigm of the late nineteenth century, now known as evolutionism.[2] Broadly, its aim was the reconstruction of human cultural development, understood as the self-evident trajectory from "simple" to "complex" forms of social organization documented in the archaeological and historical records. The ethnographic record contributed evidence of "primitive" contemporary populations, taken to represent earlier stages along the way from "savagery" to "civilization," with "advanced" European society as the end point. This use of the ethnographic data, known as the comparative method, was intended as the objective collection and sorting of facts. Any form of moral value judgment was explicitly rejected—in principle, at least. In practice, many self-proclaimed intellectuals with no credentials other than wealth and status used this approach to validate stereotypes, often biased by nationalistic interests. The myth of European superiority, with the inferiority of "primitive" societies it implied, was elevated to the status of scientific truth, typically on the basis of dubious information collected by amateur ethnologists and equally dubious standards of proof masquerading as accurate methodology (Carneiro 2003:chap. 2–5; Trigger 2006:chap. 3–6).

~34~

What started as a critique of this abuse of the ethnographic record (e.g., Boas 1896) eventually led to a reconsideration of the assumptions on which the paradigm rested. These include, for example, the preeminence of cultural parallelism over other processes, and the existence of universal standards of progress (see

---

[2] The paradigm is sometimes referred to more specifically as *classical evolutionism* or *social/cultural/sociocultural evolutionism*. These terms emphasize the distinction, both historical and conceptual, with contemporary approaches to the study of our species in light of principles derived from evolutionary biology, including evolutionary anthropology and Darwinian archaeology. The branch of contemporary approaches focusing on the process of cultural evolution (defined as change over time in the distribution of cultural traits) is also distinct from evolutionism (e.g., Boyd and Richerson 1985; Cavalli-Sforza and Feldman 1981).

Harris [2001:chap. 9–10] for discussion). Within a few decades, the paradigm had been rejected, with long-lasting repercussions for the disciplines it had given birth to. For example, key features of contemporary sociocultural anthropology can be traced back to the reaction against evolutionism in the formative decades straddling the nineteenth and twentieth centuries. Relevant features include the antipathy toward quantitative approaches and the focus on

---

If indeed there are regularities over space and time in the human phenomenon, then they must be documented in the ethnographic and archaeological records. Naturally, anthropologists are best qualified to guide attempts to extract information from these sources.   ~35~

---

field-based, site-specific investigation as the hallmark of training and practice. Comparative approaches are viewed with suspicion, even when they are completely detached, conceptually and methodologically, from the comparative method of evolutionism. More broadly, context-heavy description is preferred, and valued, over systematic explanation. Combined, these features set anthropology apart from cognate disciplines such as sociology, political science, and economics.

If indeed there are regularities over space and time in the human phenomenon, then they must be documented in the ethnographic and archaeological records. Naturally, anthropologists are best qualified to guide attempts to extract information from these sources. Yet the relative minority of anthropologists willing to engage with this research program tend to be cautious in their approach, aware that the odious excesses of evolutionism stemmed from nineteenth-century scientism. Furthermore, this minority operates among a majority who reject the research program on ideological grounds couched as methodological criticism, dismissing any scientific approach as reductionist.

Interdisciplinary research efforts continue to be hampered by this unfortunate state of affairs. My reading of the developments that led to it suggests that some caution is indeed justified, if past mistakes are to be avoided. Yet this attitude tends to frustrate researchers not familiar with the history of anthropology. As a result, the two "sides" often operate in opposition to each other, rather than in concert. The hope is that, in exposing the root cause of the tension, this brief historical sketch can lead to a more productive exchange between them.

Systematic Comparison in Anthropology

Approaches to comparative analysis in the social and historical sciences can be classified along a continuum from intensive to systematic. Intensive approaches typically involve many variables across few cases, while systematic approaches typically focus on few variables across many cases (Smith and Peregrine 2012:7–9).

To varying degrees, anthropologists are comfortable with intensive comparative approaches, generally applied informally (Trigger 2003:chap. 2). For example, it is common practice to compare and contrast societies on subsistence regime, form of social organization, and so on, to aid in interpretation of patterns and phenomena documented in the ethnographic and archaeological records. Systematic comparative approaches are more contentious, especially when coupled with formal treatment of the data (i.e., statistical analysis and/or mathematical modeling). Inevitably, there tends to be a trade-off between the number of cases and variables, on the one hand, and the amount of context (historical, ethnographic, etc.), on the other. Consequently, systematic approaches typically involve the sacrifice of detail for larger samples that are amenable to quantitative analysis. To many anthropologists the trade-off bears echoes of the comparative method of evolutionism.[3]

This attitude has stifled the application of systematic comparative approaches in anthropology throughout the twentieth century (Murdock 1971). At the same time, it has spurred methodological developments to address specific criticisms raised (see discussions of the key issues in Burton and White [1987]; Ember and Ember [2009]). One such development is the production of *standard* samples of cases, drawn from the ethnographic record, specifically for systematic comparative analysis.[4] For example, Murdock ~37~ and White (1969) collated the *Standard Cross-Cultural Sample* with the aim of adequately representing the range of cultural variation documented in the ethnographic record—that is, avoiding biases toward regions that are overrepresented. At the same time, they sought to minimize the effects of the nonindependence of human societies, arising from processes such as descent from a common ancestor and diffusion through contact[5] (Murdock 1977). Further, by establishing a standard sample, Murdock and White (1969) aimed to facilitate integration of data and findings across studies. This strategy proved successful: currently, the *Standard Cross-Cultural Sample* codebook includes coded data on approximately 2,000 variables for the 186 societies in the sample (White et al. n.d.).

In addition to these "endogenous" developments, systematic

---

[3] Indeed, the application of statistical thinking to cross-cultural samples drawn from the ethnographic record was pioneered in this context, with a paper presented by Tylor (1889) to the Royal Anthropological Institute in 1888—according to Harris (2001:158), "[p]erhaps the greatest anthropological paper of the nineteenth century."

[4] Ember and Ember (2009:93–107) provide an overview of available cross-cultural samples, standard and otherwise, including a discussion of the advantages and disadvantages of each.

[5] The issue of the nonindependence of sample units in comparative analysis was first recognized by Galton in response to Tylor's 1888 paper; Galton's comments are summarized in Tylor (1889). To this day, the issue of nonindependence is known in anthropology as "Galton's problem." It is worth noting that not all researchers agree that Galton's problem poses a threat to systematic comparative analysis in anthropology (for example, see discussion in Ember and Ember [2009:107–110]).

comparative analysis of the ethnographic record has benefited from exchanges with other disciplines. For example, since the 1970s researchers interested in the evolution of human social behavior have used this approach to uncover patterns in behavioral diversity across groups. In turn, they have contributed hypotheses (e.g., Alexander et al. 1979) and methods (e.g., Mace and Pagel 1994) from the biological sciences.

~38~ Analogous developments for systematic comparative analysis of the archaeological record have lagged behind (see discussion in Peregrine [2004]). As a result, the available resources are less known, and less used, than their ethnographic counterparts. I briefly outline the key established resources below before reviewing some challenges associated with their use. I conclude by discussing the interface with recent developments in related archaeological practice. More general overviews of comparative approaches in archaeology can be found in Peregrine (2001a, 2004) and Trigger (2003:chap. 2).

RESOURCES FOR SYSTEMATIC COMPARISON IN ARCHAEOLOGY

The major established tool for systematic comparative analysis of the archaeological record encompasses two resources developed by the Human Relations Area Files (HRAF) beginning in the late 1990s (http://hraf.yale.edu/): its online archaeological database, *eHRAF Archaeology*, and the *Encyclopedia of Prehistory* (Peregrine and Ember 2002). Both resources, described below, are used extensively across chapters in this volume (see, especially, chapters 7 and 9).

The "Archaeological Tradition" as Unit of Analysis

In an effort to address the shortcomings of previous research, development of the HRAF resources focused on the production of a standard sample of cases drawn from the archaeological record, large enough to allow for formal treatment of the data (Peregrine 2004). A key issue was defining an appropriate unit of analysis. Comparative research hinges on definition of comparable units, allowing for both generality and specificity. In archaeology, generality ensures that the definition is applicable to data from any region and time period, while specificity ensures that distinct cases remain readily distinguishable (Peregrine and Ember 2001–2002:vol. 9:2).

The HRAF resources use the "archaeological tradition" as unit of analysis. This is defined as "a group of populations sharing similar subsistence practices, technology, and forms of socio-po- ~39~ litical organization, which are spatially contiguous over a rela-tively large area and which endure temporally for a relatively long period" (Peregrine 2001b:ii).

Archaeological traditions have both a spatial and a temporal dimension: as a rule of thumb, minimal area coverage is on the order of 100,000 km², and minimal temporal duration is on the order of five centuries. The focus is on information that can be recovered from the archaeological record (e.g., subsistence practices and sociopolitical organization) as opposed to more labile traits typically used in the definition of "cultures" in ethnography (e.g., language or ideology). Consequently, an archaeological tradition may or may not correspond to a "culture" as defined for the pur-pose of comparative analysis of the ethnographic record (Peregrine and Ember 2001–2002:vol. 9:2).

The *Outline of Archaeological Traditions*

Based on the above definition, Peregrine (2001b) developed the *Outline of Archaeological Traditions* (*OAT*) as a catalog of all known archaeological traditions documenting human prehistory.

The main focus in developing the *OAT* was on extracting units roughly equivalent across areas (Peregrine and Ember 2001–2002:vol. 9:2–3), covering the entire period from the origin of the genus *Homo* in Africa approximately two million years ago to

European exploration and colonization of Oceania, the Americas, and sub-Saharan Africa approximately 500 years ago. The current version includes 289 entries (Peregrine 2001b, revised 2010).

ehRAF Archaeology

The *OAT* is the sampling frame for *ehRAF Archaeology*, HRAF's online archaeological database (http://hraf.yale.edu/products/ehraf-archaeology/). To the extent that the *OAT* is a comprehensive list of all prehistoric human societies known archaeologically (Peregrine 2004)—an assumption I discuss below—then a random sample drawn from it will be a representative snapshot of human prehistory. Based on this reasoning, *ehRAF Archaeology* provides information for a simple random sample of archaeological traditions in the *OAT*. In addition to the random sample, it provides information on complete sequences of archaeological traditions for selected world regions.

ehRAF Archaeology is continually expanding and updated annually. At the time of going to press, it covered 99 archaeological traditions overall, 48 of which are included in the random sample, with complete sequences for Egypt, Mesopotamia, the Highland and Coastal Andes, Highland Mesoamerica, the Maya area, the Mississippi River Valley, the Indus Valley, and the U.S. Southwest (C. Ember, personal communication, October 2017).

In addition to a general summary for each tradition, the database provides full-text source documents, including books, journal articles, dissertations, and manuscripts. The documents are numerically subject indexed, paragraph by paragraph, following the *Outline of Cultural Materials* (Murdock et al. 2008), a vast compendium of indexing terms that seeks to cover all aspects of human social and cultural life. This indexing system, unique to HRAF databases,[6] allows users to search for and connect related

~40~

---

[6] In addition to *ehRAF Archaeology*, HRAF develops and maintains an online ethnographic database, *ehRAF World Cultures* (http://hraf.yale.edu/products/ehraf-world-cultures/).

anthropological concepts across documents, irrespective of the language of the documents, the specific terms used, and spelling conventions. For example, a simple keyword search for "metalworking" or "smithing" would fail to retrieve related information expressed with different terms or in a language other than English. A search based on relevant subjects in the *Outline of Cultural Materials* (325: metallurgy; 326: smiths and their crafts; 327: iron and steel industry; 328: nonferrous metal industries) would instead retrieve, ideally, all related information available across documents in the database.   ~41~

The *Encyclopedia of Prehistory*

Peregrine and Ember's (2001–2002) nine-volume *Encyclopedia of Prehistory* provides descriptive information and references for 286 of the 289 archaeological traditions in the *OAT*—that is, the 286 traditions unanimously deemed "prehistoric" (Peregrine and Ember 2001–2002:vol. 9:3).

In addition to details of the archaeological record and the environment pertaining to each tradition, topics covered include the tradition's settlement pattern, economy, sociopolitical organization, religion, and expressive culture (Peregrine and Ember 2001–2002:vol. 1:x). Also included is a list of the descendants for each tradition, as determined based on time and location (Peregrine and Ember 2001–2002:vol. 9).

OUTSTANDING ISSUES

Development of a working draft of the *OAT* involved some 30 scholars over two years, called to revise and refine successive iterations of the list (Peregrine 2001b). Compilation of the *Encyclopedia of Prehistory* involved 200 scholars from 20 nations over four years (Peregrine and Ember 2001–2002:vol. 9:2–3). *eHRAF Archaeology* is a work in progress started in the late 1990s. These figures point to the impressive scale of the projects and, more generally, to the benefits of collaborative work in systematic comparative archaeology. The range of applications of the resources across chapters in this volume illustrates how they can be used to help uncover trends and patterns in human prehistory. At the same time, awareness of the challenges encountered in using these resources can prove useful in guiding future efforts (see discussion in Peregrine and Ember [2001–2002:vol. 9:1–4]). I limit discussion to two issues as they apply specifically to systematic comparative analysis of the archaeological record: derivation of a sampling frame and the statistical nonindependence of sample units.

The *OAT* is, effectively, an attempt to catalog all known prehistoric human societies (Peregrine 2004), intended as "a statistically valid sample of cases for comparative archaeological research" (Peregrine 2001a:12). But is it? One practical consideration is that, just like the ethnographic record, the archaeological record is biased. In archaeology, the bias will be toward wealthier areas and/or those with greater political stability—factors that facilitate archaeological fieldwork (Peregrine and Ember 2001–2002:vol. 9:3). Thus, to the extent that the *OAT* and the *Encyclopedia of Prehistory* provide "a snapshot of our current knowledge of the archaeological record" (Peregrine and Ember 2001–2002:vol. 9:3), they will reflect these biases, as will the random sample in *eHRAF Archaeology*.

Further, any sample drawn from these resources will comprise units that are statistically nonindependent. This can result from contact between the populations captured by different archaeological traditions, or it can occur because the populations shared a common ancestor. Both processes may lead to greater similarity between archaeological traditions that are closer geographically, for example, compared to others. Additionally, because the *OAT* is diachronic, two traditions in a sample drawn from it may represent populations that are one the direct descendant of the other. So, for example, if the earliest of these traditions developed metalworking, then it is likely that its descendant will also display metalworking. This would have to be taken into account in determining trends in,

---

The Outline of Archaeological Traditions *is, effectively, an attempt to catalog all known prehistoric human societies, intended as "a statistically valid sample of cases for comparative archaeological research." But is it?*

~43~

---

or correlates of, the acquisition of metalworking over the course of prehistory based on the sample (see discussion in Peregrine [2003]).

Accounting for the effects of all the processes described above poses nontrivial methodological challenges. Some of the issues have been discussed extensively in the ethnographic literature, as they also apply to systematic comparative analysis of the ethnographic record (see discussions in Ember and Ember [2009]; Levinson and Malone [1980]). For example, the issue of the statistical noninde-pendence of units in synchronic samples due to contact between populations or to descent from a common ancestor (i.e., Galton's problem) has attracted considerable attention, with possible recent solutions including the application of phylogenetic comparative methods (Mace and Pagel 1994) or of network autocorrelation analysis (Dow 2007). Efforts to explore how these approaches can be extended to systematic comparative analysis of the archaeolog-ical record are ongoing (P. Peregrine, personal communication, July 2015).

FUTURE DIRECTIONS

It is becoming increasingly clear that collaboration between anthropologists and data scientists will be crucial in addressing the underlying methodological issues. For example, as discussed above, the *OAT* and related resources rest on definition of a "fixed" unit of analysis (the archaeological tradition) and a "fixed" set of units (the 289 traditions in Peregrine [2001b]). With the adoption of flexible digital tools for the acquisition of data, researchers will instead be able to refine the unit they use to reflect the question at hand (see, e.g., Turchin et al. [2015] for an application to historical data).

Better still, in the future researchers may be able to bypass the a priori definition of the unit of analysis altogether. Rather, the most appropriate unit for the question at hand will be "extracted" computationally from the data. For example, data-mining and machine-learning techniques may be used to establish comparable foci of social interaction across sites, based on statistical patterns in the frequency distributions of unearthed artifacts. These techniques have been fruitfully employed in the study of other cultural domains (e.g., Michel et al. 2011), following the digitization of large bodies of data. Their application now also seems within reach in archaeology, in light of recent efforts to establish digital repositories for the preservation and some forms of integration of primary data (including the raw data and contextual information) from archaeological investigations (e.g., Open Context, https://opencontext.org, Kansa [2010]; tDAR: the Digital Archaeological Record, https://www.tdar.org/, Center for Digital Antiquity [2015]).

The aggregation and integration of both legacy and newly generated data in dedicated repositories and databanks promises an ever-changing picture of the archaeological record—a picture that will become more and more focused as the data accumulate. Although several challenges remain (Kintigh 2015; Kintigh et al. 2015), the further development of digital infrastructure in this direction is likely to transform how systematic comparative archaeology is conducted, for example extending its scope from prehistory to history (recall that the *OAT* and related resources are

~44~

restricted to prehistory). Perhaps the most important transformation will rest with how the data themselves are used. By necessity,
the typical mode of synthesis in archaeology (including any form of
comparative analysis) relies on interpretations of the primary data

---

*Discipline-wide efforts toward the development of dig-
ital infrastructure will be a crucial step in addressing
archaeology's grand challenges—fundamental questions about the human phenomenon whose answers ~45~
require information on "facts of the past."*

---

by the original investigators, or even summaries of these interpretations by others (Kintigh et al. 2015). Interpretations and summaries several steps removed from the data can become entrenched in
the literature as "facts," serving as the basis for subsequent work by
archaeologists and researchers in other disciplines. However, they
cannot be refined as more data or improved inferential procedures
become available. By contrast, the ability to access and analyze the
primary data directly will remove the need to rely on often out-
dated, or even flawed, interpretations and summaries, eventually
leading to reassessment of erroneous "facts" in the literature (Atici
et al. 2013). Additionally, the data will be more readily shared with
researchers in other disciplines and combined with complementary
sources of information, such as ecological data (Kintigh 2006).

More broadly, discipline-wide efforts toward the development of digital infrastructure will be a crucial step in addressing
archaeology's grand challenges—fundamental questions about the
human phenomenon whose answers require information on "facts
of the past," such as long-term cultural dynamics or the interplay
between ecological and social factors (Kintigh et al. 2014). For
example, why, and how, do leaders emerge in some societies, and

what sustains inequality in the long term? What drives the decline and eventual demise of societies? And how do societies respond to rapid environmental change? Tackling these and related questions will involve both synthetic work within archaeology and interdisciplinary collaboration, entailing substantial practical and intellectual challenges (Kintigh et al. 2015). The reward will be the ability to contribute to contemporary scientific and societal debates. ✦

~46~

ACKNOWLEDGMENTS
I thank Carol Ember and Peter Peregrine for feedback.

REFERENCES CITED

Alexander, R. D., J. L. Hoogland, R. D. Howard, K. M. Noonan, and P. W. Sherman
1979   Sexual Dimorphisms and Breeding Systems in Pinnipeds, Ungulates, Primates, and Humans. In *Evolutionary Biology and Human Social Behavior: An Anthropological Perspective*, edited by N. A. Chagnon and W. Irons, pp. 402–435. Duxbury Press, North Scituate, Massachusetts.

Atici, L., S. W. Kansa, J. Lev-Tov, and E. C. Kansa
2013   Other People's Data: A Demonstration of the Imperative of Publishing Primary Data. *Journal of Archaeological Method and Theory* 20(4):663–681.

Boas, F.
1896   The Limitations of the Comparative Method of Anthropology. *Science* 4(103):901–908.

Boyd, R., and P. J. Richerson
1985   *Culture and the Evolutionary Process*. The University of Chicago Press, Chicago.

Burton, M. L., and D. R. White
1987   Cross-Cultural Surveys Today. *Annual Review of Anthropology* 16(1):143–160.

Carneiro, R. L.
2003   *Evolutionism in Cultural Anthropology: A Critical History.*
Westview Press, Boulder, Colorado.

Cavalli-Sforza, L. L., and M. W. Feldman
1981   *Cultural Transmission and Evolution: A Quantitative
Approach.* Princeton University Press, Princeton, New Jersey.

Center for Digital Antiquity
2015   tDAR: the Digital Archaeological Record. Accessed June
22, 2015, from https://www.tdar.org/.

Dow, M. M.
2007   Galton's Problem as Multiple Network Autocorrelation
Effects: Cultural Trait Transmission and Ecological Constraint. *Cross-
Cultural Research* 41(4):336–363.

Ember, C. R., and M. Ember
2009   *Cross-Cultural Research Methods.* 2nd ed. AltaMira Press,
Lanham, Maryland.

Harris, M.
2001   *The Rise of Anthropological Theory: A History of Theories of
Culture.* Updated edition. AltaMira Press, Walnut Creek, California.

Kansa
2010   Open Context in Context: Cyberinfrastructure and
Distributed Approaches to Publish and Preserve Archaeological Data.
The SAA Archaeological Record 10(5):12–16.

Kintigh, K. W.
2006   The Promise and Challenge of Archaeological Data
Integration. *American Antiquity* 71(3):567–578.
2015   Extracting Information from Archaeological Texts. *Open
Archaeology* 1(1). DOI:http://dx.doi.org/10.1515/opar-2015-0004.

Kintigh, K. W., J. H. Altschul, M. C. Beaudry, R. D. Drennan, A.
P. Kinzig, T. A. Kohler, W. F. Limp, H. D. G. Maschner, W. K.
Michener, T. R. Pauketat, P. N. Peregrine, J. A. Sabloff, T. J.
Wilkinson, H. T. Wright, and M. A. Zeder
2014   Grand Challenges for Archaeology. *American Antiquity*
79(1):5–24.

Kintigh, K. W., J. H. Altschul, A. P. Kinzig, W. F. Limp, W. K. Michener,
J. A. Sabloff, E. J. Hackett, T. A. Kohler, B. Ludäscher, and C. A.
Lynch
2015   Cultural Dynamics, Deep Time, and Data: Planning
Cyberinfrastructure Investments for Archaeology. *Advances in
Archaeological Practice* 3(1):1–15.

Levinson, D., and M. J. Malone

1980   *Toward Explaining Human Culture: A Critical Review of the Findings of Worldwide Cross-Cultural Research*. HRAF Press, New Haven, Connecticut.

Mace, R., and M. Pagel

1994   The Comparative Method in Anthropology. *Current Anthropology* 35(5):549–564.

Michel, J.-B., Y. K. Shen, A. P. Aiden, A. Veres, M. K. Gray, The Google Books Team, J. P. Pickett, D. Hoiberg, D. Clancy, P. Norvig, J. Orwant, S. Pinker, M. A. Nowak, and E. L. Aiden

2011   Quantitative Analysis of Culture Using Millions of Digitized Books. *Science* 331(6014):176–182.

Murdock, G. P.

1971   Anthropology's Mythology. *Proceedings of the Royal Anthropological Institute of Great Britain and Ireland* (1971):17–24.

1977   Major Emphases in My Comparative Research. *Cross-Cultural Research* 12(4):217–221.

Murdock, G. P., C. S. Ford, A. E. Hudson, R. Kennedy, L. W. Simmons, and J. W. M. Whiting

2008   *Outline of Cultural Materials*. 6th ed. HRAF Press, New Haven, Connecticut. Revised with modifications.

Murdock, G. P., and D. R. White

1969   *Standard Cross-Cultural Sample*. *Ethnology* 8(4):329–369.

Peregrine, P. N.

2001a   Cross-Cultural Comparative Approaches in Archaeology. *Annual Review of Anthropology* 30(1):1–18.

2001b   *Outline of Archaeological Traditions*. HRAF Press, New Haven, Connecticut. Revised September 2010.

2003   *Atlas of Cultural Evolution*. *World Cultures* 14(1):2–88.

2004   Cross-Cultural Approaches in Archaeology: Comparative Ethnology, Comparative Archaeology, and Archaeoethnology. *Journal of Archaeological Research* 12(3):281–309.

Peregrine, P. N., and M. Ember (editors)

2001–2002   *Encyclopedia of Prehistory*. Kluwer Academic/Plenum Publishers, New York.

Smith, M. E., and P. Peregrine

2012   Approaches to Comparative Analysis in Archaeology. In *The Comparative Archaeology of Complex Societies*, edited by M. E. Smith, pp. 4–20. Cambridge University Press, Cambridge.

Trigger, B. G.
2003   *Understanding Early Civilizations: A Comparative Study.* Cambridge University Press, Cambridge.
2006   *A History of Archaeological Thought.* 2nd ed. Cambridge University Press, Cambridge.

Turchin, P., R. Brennan, T. E. Currie, K. C. Feeney, P. François, D. Hoyer, J. G. Manning, A. Marciniak, D. Mullins, A. Palmisano, P. N. Peregrine, E. A. L. Turner, and H. Whitehouse
2015   Seshat: The Global History Databank. *Cliodynamics* 6(1). http://www.escholarship.org/uc/item/9qx38718.

Tylor, E. B.
1889   On a Method of Investigating the Development of Institutions; Applied to Laws of Marriage and Descent. *The Journal of the Anthropological Institute of Great Britain and Ireland* 18:245–272.

White, D. R., M. L. Burton, W. T. Divale, J. P. Gray, A. Korotayev, and D. Khaltourina
n.d.   Standard Cross-Cultural Codes. Retrieved August 31, 2007, from http://eclectic.ss.uci.edu/~drwhite/courses/SCCCodes.htm.

~49~

PART II
New Research

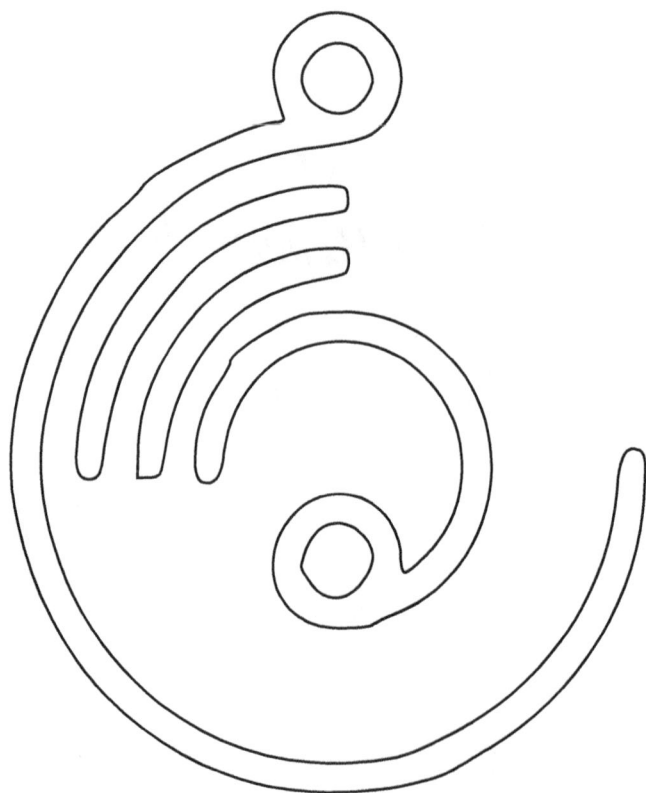

10

STATUS, ROLE, AND BEHAVIOR IN PREMODERN STATES: A COMPARATIVE ANALYSIS

Paula L.W. Sabloff, Santa Fe Institute
and Skyler Cragg, formerly Santa Fe Institute

Status-and-role, or role theory, is an old concept in anthropology that can provide new insights into the organization of premodern states. In this chapter, we adapt Linton's (1936) original concept of status-and-role to a comparative analysis of nine premodern states and contrast them with two non-state traditions in order to examine three themes of inquiry:

~53~

1. Do premodern states, many of which developed independently, organize people around similar statuses? If so, do non-states exhibit different patterns from the premodern states? Is there a pattern of statuses that suggests a society is a state?

2. Do the premodern states assign the same roles, i.e., the same rights, responsibilities and behaviors, to particular statuses—in this case, rulers, farmers, and slaves? Are these roles expected of the same statuses in non-state societies?

3. What do these patterns, if any, tell us about the organization of premodern states? And does the application of role theory help us define the premodern state?

We have found that role theory analyzed via network theory and other complexity science methods yields a productive way to understand the unique characteristics of premodern states. Some of our findings corroborate those of other archaeologists. For example, we found independent, measurable proof that Protohistoric Hawai'i fits the criteria of a state (see chapters 1 and 2 in this volume), thus supporting Robert Hommon's (2013) and Patrick Kirch's (2010) reclassification of Protohistoric Hawai'i. Other findings provide new insights into the archaeological record.

The multiple roles expected of a ruler reveal how decision-making and legitimacy operate in premodern states. The development of societal complexity means the differentiation of property rights and employment for farmers, serfs, and slaves. The many types of slavery, including indentured servitude, suggest that perhaps they should not be treated as one demographic category. And there is a strong connection between warfare and the marriage alliances of people of high rank.

~54~ Before proceeding with the analysis, we modify Ralph Linton's status-and-role theory to make it useful for the study of archaeological and complex society data.

Refining Status-and-Role for Archaeological Use

The introduction of status-and-role theory in the social sciences is often credited to American anthropologist Ralph Linton, who set forth the concept in his 1936 book, *The Study of Man.* Linton did not originate the concept, however. G. H. Mead, G. Simmel, and others started using it in the beginning of the twentieth century (Biddle 1986). Linton's conceptualization of what was to become role theory has been so efficacious for structuring research that scholars have continued to use it even as they modified his ideas (Merton 1957; Goode 1960; Goodenough 1969; Sailer 1978; Biddle 1986; Lopata 1995; Blau 1995).

Linton defines *status* as a position in society (1936:118–119). It is one level of abstraction away from an actual person. While individuals may hold certain statuses, statuses stand independent of individuals. That is, many people may occupy the same status, and a status may persist longer than the individuals who occupy it at any point in time. A status is usually expressed as a noun, one that refers to a position in the political economy (e.g., ruler, farmer, slave, or merchant), family structure (e.g., father or daughter), or some other domain such as religion (e.g., shaman, priest). Therefore, the term *status* is used as either the sum of all the offices a person occupies or just one of those offices. The status of ruler may refer to all of the ruler's positions (heir of the previous ruler, intermediary between the people and the gods, warrior, etc.), and it may mean the top position in a society—ruler of a polity. But what is important is that a status exists in relation to at least one other status. There cannot be a ruler without subjects, and a man cannot be a father without his child (biological or social).

Linton distinguishes two means of obtaining a status: *ascribed* and *achieved*. Ascribed status is assigned at birth without consideration of the individual's inherent abilities, as in the case of one's family position (e.g., son, daughter, or niece). In other cases, it may be automatically assigned according to some characteristic over which a person has no control (e.g., infant to child to youth, etc.) (Linton 1936:118–121; Ames 2008). Achieved status, by contrast, is acquired during one's lifetime. It is the result of one's efforts, often the acquisition of some skill through education or experience (e.g., doctor, Olympic medalist, criminal).   ~55~

Any given status is associated with *roles*, which Linton defines as a collection of rights and duties (1936:118–121). In premodern states, farmers have the *right* to own land and the *duty* to pay taxes, roles that farmers in non-state societies do not share. The group of roles associated with a status forms a role cluster (Merton 1957; Blau 1995). It is appropriate and expected of premodern state kings to beget an heir, form alliances through gift and marriage exchange, perform rituals for the gods, etc. These roles are usually determined by people who are in complementary positions to the status in question. The roles of the above king are influenced (but not necessarily determined) by the gods he reports to, his followers and extended family, and his subjects (Sailer 1978).

Later social scientists modified Linton's concept of role to include actual behaviors of people in particular statuses. That is, they no longer distinguished between expected roles and actual behaviors, thus simplifying Linton's ideas. Did kings form alliances because they were expected to or because they wanted to? The distinction now seemed irrelevant.

Just as status is expressed as a noun, roles and behaviors usually take the form of a verb, or action, such as *farms, inherits,* or *rules.* In each language, there are unavoidably a few instances when a role cannot be denoted solely as a verb. In this chapter, for example, we could not find a verb substitute in English for the phrase "is responsible for," a relatively common role action (e.g., the Egyptian king is responsible for the well-being of his people both in their lifetimes and after their deaths).

These basic definitions of status, role, and behavior are so much a part of the social-science vocabulary that authors no longer bother to reference the literature. Surely the concept can be useful to the analysis of premodern states, despite the existence of certain obstacles.

The first problem, of course, is that they are difficult to extract from the archaeological record. Which statuses comprise the ruler's status set? How does the agricultural labor of a farmer who owns his own land differ from that of a landless serf? Who influences the farmer's role set? We cannot answer these questions from excavation alone. But by combining archaeological and historical data, we can isolate many statuses and even their role sets in premodern states. This is the path that we have taken in our analysis. Even so, the historical data focus on the ruler and upper echelon; information on the life of slaves and even farmers is hard to come by. Therefore, there are many blanks in our database.

The second problem is that Linton's and subsequent scholars' definition of status needs to be expanded in order to be useful for understanding premodern states or any complex society. The original definition's most egregious omission is social rank. Linton defines social rank as the relative positioning of people within a family but does not systematically relate it to politico-economic rank status (1936:123, 125, 185). He connects social rank with societal hierarchy, writing that the upper echelon of a society has "privilege and rank" (125, 166) and that Comanche warriors' rank is "fluid" (446). But he does not systematically connect rank with society-level status.

We soon realized that constructing lists of titles, or positions, for a society would not capture the fact that in complex societies, an individual's statuses are embedded in a social rank. People sharing some statuses may have different ranks within a society: a land-owner may be a ruler, a commoner, or even a temple or shrine. Yet the roles associated with the status of landowner undoubtedly vary according to the status holder's rank in the society: landowning rulers assign to others the task of farming their land, while land-owning commoners usually farm it themselves.

~57~

We also found that Linton's distinction between ascribed and achieved status is inadequate for complex societies, and we have therefore added another category—*imposed* status—to refine the analysis. We define imposed status as one that is forced upon a person and is therefore neither ascribed nor achieved. Once a war-rior or soldier is captured and becomes a prisoner of war, he may be forced to become a sacrificial victim or slave, while a woman who is made to marry has the status of wife imposed on her.

When we studied the statuses of slave, captive, and human sacrifice, we also realized that performing a role or behavior may be fulfilled in two manners. A role/behavior may be active, that is, it is performed under the person's agency or initiative. But a role/behavior may also be passive. Here, a person is forced or coerced to undertake a task. This distinction can often differ greatly depending upon the particular status-rank combination under scrutiny. Indeed, it appears that any role cluster varies in its mix of active and passive roles depending on the rank of the person in a particular status. We characterize the roles associated with statuses from ruler to slave as lining up along a continuum from mostly active to mostly passive.

Methodology and Cases

We have been building a database of different traditions' statuses and roles, from captives and slaves to rulers, gods, and spirits. Researchers working with us have pored through the archaeological and historical literature to construct the database. Their work, which originally appeared as citations from the literature, was checked by an expert on each tradition.

Table 1 is an abbreviated example of the categories used in the societal databases. The information is organized into rows: a tradition's location and temporal spread; the names of all the statuses we could find (in English and the original transliterated language when possible); and the expected roles or actual behaviors associated with each status. To determine rank, we recorded each status's place in the sociopolitical hierarchy and included a "Reports to" row. These turned out to be difficult to ascertain, given that an occupational status could be held by people with very different ranks. A diviner in Late Shang China might be a member of the royal family or a commoner. Following a status's place in the hierarchy is a row devoted to how an individual might legitimately attain that status. We reserved the last row for comments, or information the researcher felt would shed light on the status but which did not fit into one of the preceding rows. Usually these are insights into the context of the status or roles.

We found that we needed historical or epigraphic reports to flesh out the archaeological data, and in most cases writing did not crystallize until the end of the premodern state period. Therefore,

TABLE 1 Example of how the data are organized in each society: Shang China.

Associated territory/dates	(Early–Late) Shang China 1600–1050 BCE
Status	King (*Wang*) (Trigger 2003:89)
Expected roles	• Rules over a state that covers most of north-central China (Campbell 2015:3) • Travels around the territory (Keightley 1983:537) • Delegates authority to territorial governors (Trigger 2003:216) • Accepts tribute from lords living outside the capital (Barnes 1999:134)
Place in hierarchy	Top of living beings, but gains divine ancestral status upon death
Reports to	Royal ancestors, chief god
Basis of legitimacy	Born into ruling clan; oldest son or next oldest brother of king (Feng 2013:103–106; Trigger 2003:149–150)
Comments	King was not considered "the son of heaven"—the representative of the gods on earth—until the Zhou dynasty (A. D. Smith, personal communication, 2014)

although we had hoped to capture the initial state period, we had to use the period in which writing had developed. Furthermore, today's scholars had to be able to decipher the written record. For example, while the state formed in Egypt when Narmer united the Lower and Upper Nile (the Early Dynastic Period, 3150–2686 BCE), we use as our reference point the Old Kingdom (2686–2181 BCE), the period when the hieroglyphic system existed and could be read by today's scholars. Similarly, although the Indus Valley has written records, scholars still cannot read them, and we could not use this tradition in our analysis.

Primary Premodern States

Our selection of cases encompasses all the primary premodern states: that is, those that self-organized into more complex organizations without using a model developed in another state:

- Late Shang Dynasty China (ca. 1250–1046 BCE), analyzed by Paula L.W. Sabloff and checked by Adam Daniel Smith.
- Old Kingdom Egypt (2686–2182 BCE), analyzed by Robert S. Weiner and checked by Laurel Bestock.
- Protohistoric Hawai'i (1650–1778 CE), analyzed by Kong Fai Cheong and checked by Patrick Kirch.
- Early Dynasty III, Lower Mesopotamia (2600–2350 BCE), analyzed by Henry Wright and checked by Stephen Tinney.
- Late Preclassic to Terminal Classic Maya (400 BCE–900 CE), analyzed by Kong Fai Cheong and checked by Jeremy Sabloff.

Because of the need to use texts as well as archaeological remains, the time periods for the states in our sample usually date several generations after the actual emergence of the states. In the case of Protohistoric Hawai'i, the time period used is the one reported on by the British.

Secondary Premodern States

The secondary premodern states in our analysis were stimulated by earlier states' complex social structure. For example, Mycenaean Greece had contact with pharaonic Egypt and Minoan Crete (Dickinson 1999). We selected four from different geographic areas based on the suggestions of our advisors:

- Aztec (1350–1520 CE), analyzed by Jeffrey Cohen and checked by Michael E. Smith.
- Benin (ca. 355 BCE–ca. 1092 CE), analyzed by George J. Haddad and checked by Sandra Barnes.
- Mycenaean Greece (1300–1200 BCE), analyzed by Skyler Cragg and checked by Michael Galaty.
- Postclassic Zapotec (1050–1500 CE), analyzed by Kong Fai Cheong and checked by Gary Feinman.

Non-States

For comparison with the state traditions above, we used two non-states. These were chosen because the Santa Fe Institute had several experts in the area and therefore could guide our research.

- Tewa, preconquest (1300–1600 CE), analyzed by Jack Jackson and checked by John Ware.
- Hopi, the ethnographic present from the first half of the twentieth century, analyzed by Jack Jackson and checked by Abigail Holman.

Although we have continued to add more cases to the database, we limit this chapter to the above, which have been checked by experts. Following sign-off from all experts, we compiled the data from all 11 societies into one spreadsheet listing statuses (rows) and expected roles and actual behaviors (columns) and reducing each cell to a 1 (present), 0 (not present), or blank (missing information). In addition to the information from the database, we included information deduced from logic. For example, we assumed that a ruler did not normally become a human sacrifice or perform corvée labor in his/her own polity. We assumed that since royalty did not pay taxes, the ruler did not either. Be that as it may, we tried to err on the side of caution. If the literature did not mention slaves, we left the cell blank. If an author reported that there is no evidence of slaves, we placed a 0 in the appropriate cell.

~61~

We imported the database into UCINET 6 (Borgatti, Everett, and Freeman 2002), performing a Johnson's hierarchical cluster analysis to find the core attributes of each network. Generated from a symmetric (square) matrix, Johnson's hierarchical cluster groups together the pairs of attributes that share some characteristic. Here, the shared feature is the relative frequency of the pairs' co-occurrences, which are derived from the weighted average and are represented as dendrograms in Figures 2, 4, and 6.

We also created graphs in NetDraw (Borgatti 2002) to illustrate the co-occurrence of roles/behaviors for the statuses of farmers/

herders, serfs, and slaves. In these graphs, the edges represent co-oc-currence between pairs of nodes.

We also ran categorical core-periphery analyses in order to have an independent check on the results of the cluster analyses. Both tests were generated in UCINET 6 (Borgatti, Everett, and Freeman 2002; Hanneman and Riddle n.d.; see also Brughmans 2013).

Statuses in Premodern States

We gleaned 52 possible statuses in premodern states from the literature (Table 2). They range from "god" and "ruler" to "captive" and "slave." We first investigated which statuses form a "core" in the premodern states—that is, which statuses are found in most (if not all) premodern states yet may not be present in the non-state cases. We performed a Johnson's Hierarchical Clustering analysis on the nine (five primary and four secondary) states and found that most shared 33 statuses. The non-states exhibited only 10 of the core statuses and were therefore not included.

We then divided the core statuses by domain and social rank (Table 3). Domains are major categories of behavior. Here, we use four. The sociopolitical domain ("SP" in the third column) refers to class or political statuses such as royal/noble or commoner and ruler or provincial governor. It was difficult to distinguish between the two for these early complex societies. The economic domain ("E") covers statuses associated with work or productivity, e.g., farmer, while the religious domain ("R") includes statuses of people or beings affiliated with the supernatural. The external interaction domain concerns statuses associated with premodern states' contacts with other polities. Because the main interaction in this group is warfare, we have marked this domain "W."

We also grouped the statuses by rank, noting that some can be filled by people of more than one hierarchical level. Rank 1 in the fourth column of Table 3 is reserved for supernatural beings, all of whom are found in the R (religion) domain. Rank 2 is the highest rank a human mortal (not an ancestor) may attain. Such statuses range from ruler to his wife or concubine and then royal/noble. High-ranking political statuses range from ruler (usually ruling by divine right) to territorial official.

Below these ranks is the commoner (3), who is sometimes called freeborn (see Bradbury 1973:60 on Benin) or a member of the free class in Lower Mesopotamia (Gelb 1965:241). A commoner may serve as a low-level official or bureaucrat. ~63~

The lowest rank (4) is assigned to a slave or captive who acts as a servant, laborer (skilled and unskilled), and sometimes supervisor of laborers. A slave may have been a captive or criminal, as in the Aztec case (M. E. Smith 2003:137). Slaves' and captives' statuses are imposed, and their lives are controlled by another agent, usually someone from rank 2. Slaves work but do not earn (Trigger 2003:157–160).

We found that non-states have neither the top nor the bottom ranks found in premodern states. Instead, all people are considered to have similar rank, and any gradation among them is not hereditary. Non-states have the concept of *the people*, which is frequently the name they call their group (e.g., *Diné* in Navajo means "the people"). In contrast, those living in premodern states are part of a stratum within a larger society.

TABLE 2 All possible statuses found in archaeological and historical literature on premodern states and non-states.[1]

	Status	Obtained	Definition of status
1	Ruler of rulers (para-mount ruler)	As and/or Ach	Ruler of other rulers and their territories as well as one's own. A paramount ruler.
2	Ruler	As and/or Ach	Ruler of a territory or people.
3	Queen ruler	As and/or Ach	Female ruler who controls/rules in her own right or as regent.
4	Prince	As	Son of a ruler; possible heir to his father's position.
5	Ruler's female kin	As	Daughter, sister, etc., of a ruler. While her natal position is ascribed, her roy-al kin may impose her adult status.
6	Queen con-sort/mother	Ach or Im	Senior wife of a king or paramount ruler.
7	Ruler's other wife	Im	Attached to the ruler through mar-riage or some facsimile of marriage.
8	Royalty/no-bility (male)	As	A royal is a person with direct kinship ties to a ruler. A society determines how many generations out to count royalty. A no-ble is a person whose kin bonds to royal ancestors are too distant to be counted as royal but is still part of the elite.
9	Royalty/no-bility (female)	As and/or Im	Royal or noble woman, depending on her lineage or her marriage.
10	Provincial elite	As	A person who was part of the ruling stra-tum of a territory before it was conquered.
11	Commoner	As	A person not born into a royal or noble lineage. Some societies include serfs in this category; others do not. It varies in the literature. A female commoner is born or marries into a commoner lineage. She usually ranks a step below her husband.
12	Commoner leader	As or Ach	Within the commoner stratum, he is head of a lineage and is responsible for the behavior of its members.
13	Serf	As	A person who farms, herds, or does general work for an overlord. He does not own land.

~64~

---

[1] How a particular status is obtained is represented in the third column by "As" (ascribed by birth or a change from one stage of life to another), "Ach" (achieved through per-sonal effort), or "Im" (imposed by another person or derived from a preceding status, for example, an enemy soldier can be made a captive and then a slave).

TABLE 2 (*continued*)

	Status	Obtained	Definition of status
14	Slave	As or Im	Someone who works for an overlord or some corporate entity, e.g., the state, the ruler, or a temple/shrine. He/she is the owner's property. A slave has no freedom of movement; he/she receives compensation (food, clothing, etc.) rather than a wage.
15	Captive	Im	Someone from outside the polity who has been captured (usually) on the battlefield or by raiding within the enemy's territory.
16	Human sacrifice	Im	A person who is destined for sacrifice. He/she may be a prisoner of war, a slave, etc.
17	Prime minister	As and Ach	A person who advises the ruler or makes policy decisions.
18	Council member	As and Ach	A member of a council that advises the ruler or makes executive decisions within a circumscribed territory. He/she operates on several levels of government.
19	Provincial governor	As and Ach	A person who controls a captured territory or people in the name of the ruler. He is usually a member of the royal family or the nobility, but sometimes he is the former ruler of the captured territory.
20	Territorial official	As and Ach	Someone who is responsible for a cluster of settlements within a ruler's territory that is smaller than a province.
21	Settlement/ ward official	As and/ or Ach	A person who governs a settlement or part of a settlement, e.g., a town, village, or hamlet.
22	Government administrator	As and Ach	An official responsible for overseeing a part of the bureaucracy, e.g., agriculture, treasury, public/religious works. Or a midlevel administrator who supervises a section within a department such as foreign relations, agriculture, or tax collection. He may also be a tax collector.
23	Scribe	As and/ or Ach	Someone who writes accounts. Such writings may be divinations or other religious accounts, tax rolls, etc. He/she may be a royal, noble, or commoner.
24	Military commander/ warrior	As and Ach	A general or war chief who decides military strategy or leads an army. He/she may be anyone from a ruler to a member of the nobility. In Benin, he may sometimes be a slave.

TABLE 2 (*continued*)

	Status	Obtained	Definition of status
25	Age-grade member	As and Ach	When a society is divided into age grades, each male above puberty is initiated into an age-grade of his cohorts. Together they move through the grades of youth, adult, and elder.
26	Fief holder	As and/ or Ach	Someone—usually a royal or noble—who owns swaths of productive land (agricultural or pastoral); a landed aristocrat.
27	Retainer	Ach, perhaps As	Someone who directly serves the ruler, his/her immediate family, or other members of the court. Not a slave, he/she often comes from the royal to the commoner stratum.
28	Supervisor (skilled workers)	Ach	Someone who oversees the productivity of skilled workers, e.g., craftspeople, entertainment groups, masons, etc.
29	Supervisor (unskilled workers)	Ach	Someone who oversees the productivity of unskilled workers, e.g., ditch diggers or field hands.
30	Skilled worker	As, Ach, or Im	An artisan or craftsman. Someone who makes and/or repairs anything from chariots to ceramics. Often there is a hierarchy within this category, and the master craftsman may run a workshop or train apprentices.
31	Unskilled worker	As or Im	Someone who performs unskilled labor, e.g., farming, ditch digging, or carrying bricks.
32	State worker	As, Ach, or Im	Someone who works for the ruler, the state, or a temple.
33	Private worker	As, Ach, or Im	Someone who works for a private overlord.
34	Farmer/ herder	As and/ or Ach	Someone who tills the soil and/or raises animals for a living (a herder or shepherd). The land may or may not belong to him/her. Often he/she performs corvée labor and is a commoner.
35	Soldier, police	Ach and/ or Im	Someone who serves full-time in the ruler's army, guard, or police. While farmers often serve as foot soldiers in times of conflict, soldiers who drive chariots, archers, etc., may be full-time.
36	Peddler/ trader	As and/ or Ach	Someone who exchanges goods and produce on a local level. The peddler or trader may operate within a barter or market economy.
37	Merchant	As and/ or Ach	Someone who exchanges goods or produce usually for luxury items or resources coming from outside the state or from long distances within the state. The merchant may operate within a barter or market economy.

TABLE 2 (*continued*)

	Status	Obtained	Definition of status
38	Servant	As and/ or Ach	Someone who works for another and receives wages.
39	State priest	As and Ach	The high priest for the entire state. He/she is often the ruler or some close kin of the ruler.
40	Head priest	As and/ or Ach	The high priest/priestess for a shrine or temple. He/she may be some close kin of the ruler.
41	Servant of the gods	As and/ or Ach	The person(s) responsible for making sure the gods are "fed" through sacrifice or offerings. He/she may be the ruler.
42	Priest, priestess	As and/ or Ach	A person responsible for religious activities perhaps in a temple or shrine. He/she may perform specialized rituals, e.g., sacrificing animals or humans, performing ceremonies over the dead, and prewar rituals.
43	Temple admi-nistra-tor/staff	As and Ach	The chief administrator directs others in maintaining shrines, temples, and the ritual paraphernalia used in rituals.
44	Religious specialist	Ach	A performer for religious rites—music, dance, etc.
45	Diviner	As and/ or Ach	Someone who predicts the future or assists another in ceremonies to predict the future.
46	Shaman/ magician	As and/ or Ach	Someone who connects individuals or groups to the spirit world, often for curing purposes. One who uses magic to accomplish a task such as healing, cursing, or killing another person. He/she may operate outside the religious hierarchy.
47	Chief god		The high god, usually the creator of the world or the progenitor of other deities.
48	Lesser god		Any deity ranked below the chief god.
49	Nature spirit/god		A deity that controls nature—rain, sun, agriculture. He/she may also be a lesser god.
50	Settlement god		The chief deity for a settlement.
51	Ruler's an-cestor spirit		The spirit of a dead ancestor who was in direct line to the current ruler.
52	Others' an-cestor spirit		The spirit of a dead ancestor of others—elites and non-elites alike.

~67~

TABLE 3 (*opposite*) Core statuses of premodern states by domain, rank, and society. Domains in the third column include sociopolitical (SP), economic (E), religious (R), and external interaction (W). Rank ranges from 1 (deities) to 2 (ruler to nobility) to 3 (commoner, i.e., neither noble nor slave) to 4 (slave, captive). The numbers in column 5 are the percent of premodern states (Egypt, Lower Mesopotamia, China, Maya, Hawai'i, Aztec, Zapotec, Greece, and Benin) that are known to have that status. For the Tewa and Hopi, 0 means the status is not present; 1 means the status is found in the literature; and empty spaces mean the data are missing.

CORE STATUSES OF PREMODERN STATES
BY DOMAIN AND SOCIAL RANK

Certain features stand out when we look at the core statuses by domain (major categories or institutions of behavior) and social rank. The top eight core statuses in the sociopolitical domain may be occupied only by people of the highest rank (rank 2). They are: ruler, prince, ruler's female kin, queen consort/mother, royalty/ nobility (man), royalty/nobility (woman), provincial elite, and territorial official. Four more statuses found in the sociopolitical domain draw members from either the second (upper echelon) or third (commoner) rank. These are retainers to the king or other royals, secondary wife or concubine of the ruler, settlement or ward official, and government administrator. The duties and territorial responsibilities of the latter vary with rank.

By combining rank and status in the domain, we are made aware that certain positions, or offices, must be occupied only by members of the upper echelon while others are not. This is generally true for military statuses as well.

The agrarian sector of the economic domain (E) cross-cuts rankings. For example, the highest-ranked status concerned with farming is fief holder, that is, a royal or noble who receives income from estates.[1] Commoners (rank 3) engaged in agriculture may be

---

[1] An exception is Old Kingdom Egypt, where the early rulers "own" all arable land. Royal and noble followers administer the estates and receive income from them but do not own them in the legal sense (Blanton and Fargher 2008:147; Baines and Yoffee 1998:229).

	Status	Domain	Rank	States	Tewa	Hopi
1	Ruler (king or queen)	SP	2	100%	0	0
2	Prince	SP	2	100	0	0
3	Ruler's female kin	SP	2	100	0	0
4	Queen consort/mother	SP	2	100	0	0
5	Royalty/nobility (man)	SP	2	100	0	0
6	Royalty/nobility (woman)	SP	2	100	0	0
7	Provincial elite	SP	2	89	0	0
8	Territorial official	SP	2	89	0	0
9	Retainer	SP	2, 3	100	0	0
10	Ruler's other wife	SP	2, 3	100	0	0
11	Settlement/ward official	SP	2, 3	100	1	1
12	Government administrator	SP	2, 3	100	0	0
13	Commoner	SP	3	100	1	1
14	Slave	SP	4	100	0	0
15	Captive	SP	4	100	0	0
16	Fief holder	E	2	89	0	0
17	Supervisor of skilled workers	E	2, 3	89		
18	Peddler/trader	E	3	78		
19	Farmer/herder	E	3	100	1	1
20	Serf	E	3	100	0	0
21	Skilled worker	E	2, 3, 4	100	1	1
22	Supervisor of un-skilled workers	E	3, 4	100	1	1
23	Unskilled worker	E	3, 4	100	1	1
24	State worker	E	3, 4	100	0	0
25	Private worker	E	3, 4	100	1	1
26	Servant	E	3, 4	78	0	0
27	Military commander/warrior	W	2	100	1	1
28	Soldier, police	W	3	100		
29	Servant of the gods	R	2	100	0	0
30	Priest/priestess	R	2, 3	89	0	0
31	Chief god	R	1	100	0	1
32	Lesser god	R	1	89		
33	Nature spirit/god	R	1	89	1	1

farmers, herders, or shepherds. Serfs' civil status varies between commoner and slave (ranks 3 and 4).

The ranks of those engaged in farming illustrate that people in different ranks but engaged in the same activity may have very different property relations. In the premodern states, rulers, fief holders, and sometimes temples or shrines own swaths of arable land; farmers own some land; and serfs and slaves own no land but work for an overlord. In the non-states, farmers all have the same ownership relation to the land. They either own the land outright or have usufruct rights to some of the land owned by their settlement or extended kin group (McAnany 1995:92–96). Indirect evidence suggests that these practices were established in Pueblo societies in the first millennium CE (Kohler 1992).

All the premodern states institutionalize different kinds of property rights, from royal ownership to fiefdoms, corporate (kin or settlement group) landholding, and small plots. Thus the real import of agriculture for state formation is not dietary change or sedentarism but diversification in types of productive property linked with the right to hereditary property, or land. This transition preceded state formation (see Johnson and Earle 2000:201–202; Flannery and Marcus 2012:206–207, 256–257).

In the industrial sector of the economic domain, we see a parallel process of diversification of craft production by rank. Sometimes even royals make fine crafts such as Mayan polychrome vases (Houston and Inomata 2010:263) or act as diviners, as did some royals in Late Shang China (Feng 2013:93, 108; Trigger 2003:505; A. D. Smith 2010, 2011). However, we do not have evidence of royals or nobles engaged in unskilled labor, which is universally left for commoners and slaves.

Denizens of states may also work for different types of employers. In Lower Mesopotamia, craftsmen may work for themselves, selling their own wares, or for a private person, such as a royal, a noble, or a kinsman. They may also work directly for the state. In Mycenaean Greece, skilled workers may work for a temple or the state (A. Westenholz 2002:30; Gelb 1965:242). In Old Kingdom Egypt, they work in teams administered by controllers and various overseers (Baines and Yoffee 1998:230; Trigger 2003:369). In Late Shang China, makers of bronze vessels work for royal or noble patrons, usually in workshops (A. D. Smith 2010; Campbell 2015; Campbell, Li, He, and Jing 2011).

Differentiation of possible employers seems to be another key feature of state societies. A role-theory analysis of a population ~71~ helps illuminate this fact.

In the religious domain (R), premodern states have a hierarchy of deities, from a chief god to lesser gods, gods of particular settlements, nature spirits, and sometimes ancestor spirits (found in Late Shang China, Classic Maya, Benin, Postclassic Zapotec, and Protohistoric Hawai'i). This ranked set of deities reflects and perhaps legitimizes the sociopolitical hierarchy: just as there are at least three levels of society (ruler and nobility, commoner, and slave), so there are at least three possible levels of gods.

Humans working in the religious domain also belong to several ranks. The Aztec, Postclassic Zapotec, and Protohistoric Hawaiian states have a state priest (a priest over the entire state), specialized priests, and regular priests. The Aztec also have head priests for different temples and shrines.

The status of "servant of the gods" refers to someone who is responsible for performing certain rituals to honor the gods or bringing sacrifices and offerings to the gods' statues. Often the ruler has this status and thereby possesses a direct (and often monopolistic) connection to the gods. The one who satisfies the gods' needs has the ability to please, appease, interpret, and ask favors of them.

As noted previously, we view the rankings in every domain as a progression from no decision-making power (slave) to complete freedom (chief god) and from poverty (serfs and slaves) to wealth (ruler and royalty). We do not see these continua in the Tewa and Hopi columns of Table 3. Although these societies appear to share almost half the statuses with the states, we must keep in mind that the categories were developed for state-level societies; the non-states' data were fit into categories that might not make sense for them. For example, a Hopi may have the title of military commander, but by the period of research, the Hopi no longer practiced warfare.

~72~

HIGH-RANKING WOMEN AND WAR

Within the states' sociopolitical domain is a cluster of core statuses reserved for high-ranking women other than a queen ruler. These are the (male) ruler's female kin, the queen consort/mother, and royal/noble women, whose marital statuses are imposed rather than achieved. They are core statuses probably because they have great importance for the ruler, for they provide him with long-term alliances. Unlike a bronze vessel or piece of jewelry, high-born women whom the ruler marries to a (potential) ally or women who marry into the ruler's entourage maintain contact with their kin for years, passing messages and advocating for their kin or spouse. The literature provides some examples of alliance building through the imposed intermarriage of women (Schele and Freidel 1990:59; J. G. Westenholz 1990; Freidel and Schele 1997; Keightley 1999:33, 43; Freidel and Guenter 2003; Allen and Arkush 2006:4; Connell and Silverstein 2006:400, 402; LeBlanc 2003, 2006:406; Turchin, Currie, Turner, and Gavrilets 2013, among others). Most telling is Simon Martin's (2008) description of the Dallas Altar, which depicts three high-class Mayan women from the Snake kingdom (Calakmul) who marry into the smaller settlement of La Corona. As the three come from different generations, they cement the ties between the two settlements.

Scholars who have written on the impact of warfare on pre-modern state formation mostly mention alliance building only in passing (see also Stanish and Levine 2011). The prevalence of long-term alliance maintenance—implied by the centrality of high-rank women's statuses—suggests that it must be seen as a concomitant process to warfare. Rulers and the elite need trustworthy allies in order to wage war against a mightier foe—or assure victory over any foe. Sometimes warfare is a last-resort effort that occurs only when long-term alliances break down. Therefore, we urge archae-ologists to treat long-term alliance and warfare as two sides of the same coin.

Using women to actualize long-term alliances depends on polygyny, which grants a ruler the ability to form and maintain several marriages simultaneously. While we do not yet know how far down the social ladder polygyny is practiced in many of the states we studied, we do know that it was practiced by the ruler (king) in all nine of the premodern states.

Typical Premodern States

Once we determined the core statuses, we used that informa-tion to discern whether there exists a "typical" premodern state, one that stands as an example for research or public interest. We devised a prototype from the vector of average percentage of sta-tuses along three dimensions—domains, ranks, and domains and ranks combined— and then measured the Euclidean distance of the different societies from the vector. Table 4 shows that the typ-ical society varies according to the dimension selected in a table of Euclidean distances. There is little variability between the states in the domains of external relations and the economy. That is, there is little variability in the number of statuses found in these domains. However, the sociopolitical and religious domains exhibit suffi-cient variability to make Late Shang China and the Aztec closest to the prototype. Lower Mesopotamia, Protohistoric Hawai'i, and Benin have the same number of statuses as Late Shang China and the Aztec for the sociopolitical domain, but their scores for the religious domain are not at all near the prototype. Needless to say, the two non-states vary greatly from the individual states as well as the prototype.

Regarding societal ranks, the states differ in the number and kinds of statuses of deities (and supernatural phenomena). Old Kingdom Egypt deviates the most from the others as well as from the prototype. Considering the highest rank for humans—from the ruler to the nobles—there is a greater degree of shared statuses than there is for the religious hierarchy. All the states have slaves and captives, and so there is little variability among states in this rank. When the four ranks are averaged, the Classic Maya are closest to the prototype, followed by Protohistoric Hawai'i and Benin. Again, the non-states do not share the hierarchy of people that the states do, and so the non-states' scores are much lower than the states'.

Combining the domains and rankings, Lower Mesopotamia is closest to the prototype, with the Classic Maya following close behind. In other words, Lower Mesopotamia has the lowest proportion of deviations from the prototype of all the states.

Table 4 also highlights other patterns. First, Protohistoric Hawai'i's place in all three parts of the table suggests that it has many of the characteristics of a state and therefore should be considered one. Second, the Hopi and Tewa societies are at the bottom of all three measurements. Their relatively greater distance from the prototype gives added weight to their classification as non-states, which the archaeological and ethnohistoric data bear out (see below). Third, Old Kingdom Egypt appears just above the Tewa and Hopi in the ranks-only and combined sections. We postulate that its position is the result of an ideology of regal divinity,

TABLE 4 Each society's Euclidean distance from a prototype of an early state (ranked from smallest to largest distance).

Based on domains only		Based on ranks only		Based on domains and ranks	
China	0.001	Maya	0.008	L. Mesop.	0.015
Aztec	0.001	Hawai'i	0.009	Maya	0.018
L. Mesop.	0.002	Benin	0.009	Benin	0.030
Maya	0.010	L. Mesop.	0.012	Greece	0.034
Greece	0.020	Greece	0.015	Hawai'i	0.037
Benin	0.021	Zapotec	0.056	China	0.066
Hawai'i	0.028	China	0.065	Aztec	0.076
Egypt	0.052	Aztec	0.075	Zapotec	0.120
Zapotec	0.064	Egypt	0.076	Egypt	0.128
Hopi	0.665	Hopi	0.614	Hopi	1.279
Tewa	1.022	Tewa	1.038	Tewa	2.060

that is, the principle that the Egyptian king was the son of the god Horus and therefore divine himself. The other traditions (with the exception of the last years of the Hawaiian state) legitimize the ruler by the principle of divine right—the god-given legitimacy to rule—rather than divinity.

We were also interested in which societies shared the most statuses. We were able to compare the relative number of matched (0 to 0, or 1 to 1) and mismatched (0 to 1) statuses between the societies, eliminating missing values and therefore giving a more accurate picture of what is shared. Figure 1, based on the original 52 possible statuses (see Table 2), illustrates the co-occurrence of statuses in pairs of societies. For example, it shows that in the first column, Old Kingdom Egypt and the Classic Maya have the greatest number of matches (83 percent) and therefore the fewest (17 percent) mismatches. In the second data column, Lower Mesopotamia shares 81 percent of matches with Mycenaean Greece and the Aztec, but the latter two only share 69 percent with each other. These

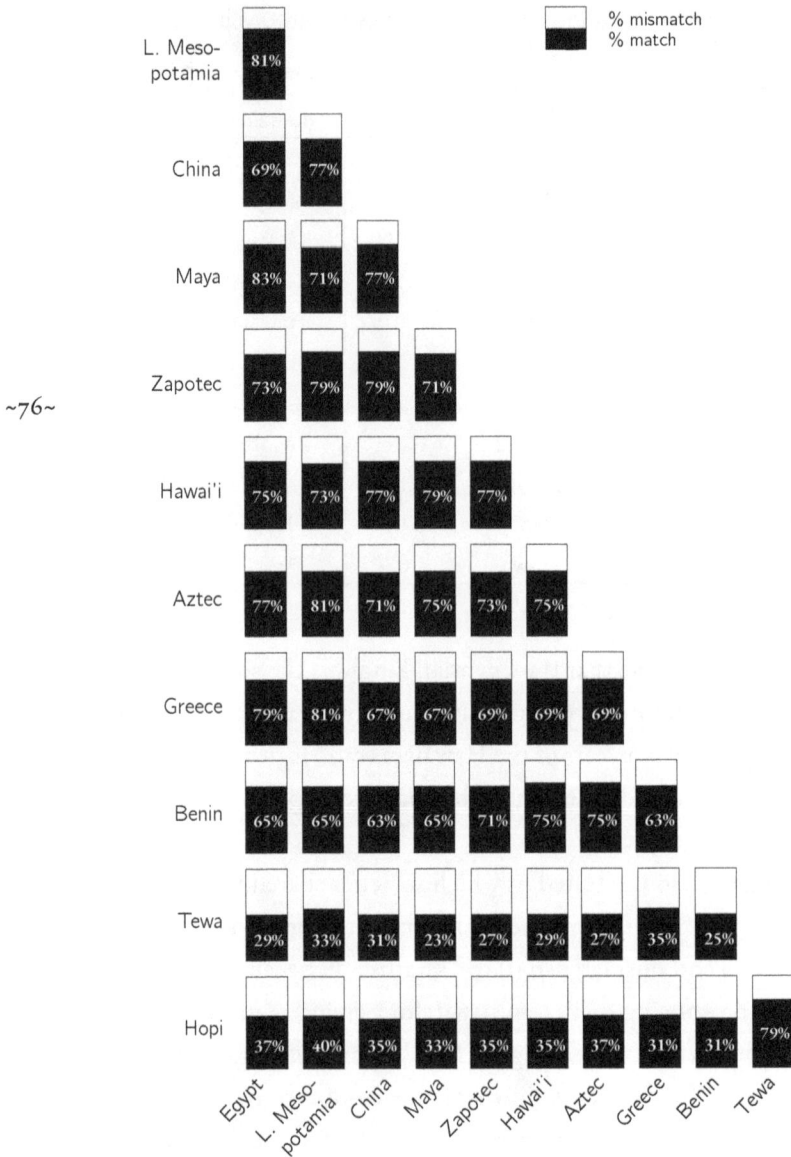

THE EMERGENCE OF PREMODERN STATES

~76~

FIGURE 1 Similarities between societies as determined from matches and mismatches on 52 statuses. For each pair of societies, the solid gray part of a bar represents the percent of the statuses that are known to match (that is, both societies share the same number of statuses). The white part of a bar represents the percent of known mismatches, such as when one society has a diviner but the other does not. Missing data have been eliminated.

intercontinental pairings suggest there is little difference in the number and kinds of statuses between state organization in the Old and New World or between primary and secondary states.

The non-states share few statuses with the premodern states. While the Tewa and Hopi share 79 percent matches with each other, the Tewa—the least hierarchical society in this sample—share only 35 percent of statuses with its next-highest percentage of matches (Mycenaean Greece). The latter is the least hierarchical of the premodern states, as far as we know. The Hopi, in turn, share only 40 percent of statuses with Lower Mesopotamia. ~77~ Kohler (2013) suggests that the Tewa and Hopi, who belong to different language groups but the same cultural tradition, may share sociocultural patterns because of migration (frequent interchange of populations and ideas) and adaptation to a similar environment.

Roles/Behaviors for Three Statuses in the Premodern States

In this section we present an analysis of the roles/behaviors for rulers, farmers, and slaves using our modified definition of Linton's concept. These statuses were selected because they represent different ranks in the social structure. Rulers are always of the nobility (rank 2), and farmers are generally from the commoner stratum (rank 3), i.e., neither noble nor slave. Slaves (rank 4) may play different roles, depending on their skills, their status before enslavement, and the status/rank of their owner.

THE RULER

The ruler, the person with the highest rank in a polity, performs more roles than any other status found in a premodern state—33 out of a possible 66. His roles also extend through all four domains of the study, as seen in the fourth column of Table 5. Thirty are core roles, for they are shared by at least eight of the states in our sample according to Johnson's hierarchical clustering analysis and the core-periphery test.

While Table 5 is meant to represent both male and female rulers, certain roles likely apply only to a king. We assume that queen rulers, who appear in the literature on Old Kingdom Egypt and the Classic Maya, do not practice polygamy, although kings in all premodern states do. (One hears of harems but not stables of husbands.) In order to form many long-term alliances, the queen ruler must depend on kin or wait for her children to be of marriageable age. Still, both king and queen rulers use patronage to tie their royal subjects and conquered territories to them.

There are 14 roles/behaviors within the ruler's sociopolitical domain. Of these, only two may be interpreted as having a passive component. "Inherits" (row 13) may appear to be passive, but a ruler often has to compete against others to inherit his position. "Predicts" (row 31) is ambiguous, as it is not always clear whether the gods are predicting and the ruler is interpreting or the ruler is doing the predicting. Therefore we marked "predicts" as both active and passive.

The roles that have both sociopolitical and economic dimensions (rows 15–19) are all active, as is row 20, which combines sociopolitical and religious functions. The five external-relations roles are by and large active, for the ruler makes alliances and wages war. Row 26 has a passive component: when the ruler is captured in war. Being captured is definitely passive, although it derives from being an active fighter.

In the religious domain (rows 27–30), the ruler may be considered passive only when he is seen to be serving the gods or ancestors.

Looking at the ruler's roles in Table 5, we gain insight into his/her position in the sociopolitical order. In essence, we see a person who controls power in several ways. By monopolizing access to the chief god or royal ancestors, the ruler gains legitimacy and

TABLE 5 Roles, domains, and agency of a premodern state ruler, including all possible roles, starting with the core roles (i.e., those that occur in at least eight of the nine states). Domains include SP (sociopolitical), E (economic), R (religious), and W (external).

	Role	Definition	Domain	Active/passive
		CORE ROLES		
1	Rules	Rules (makes policy decisions for) a territory (from a kingdom or empire to a province or village; makes rules/policy) or bureaucratic unit (e.g., government department such as treasury)	SP	A
2	Administers	Carries out policy decisions by administering a government department, public works, etc.	SP	A
3	Judges	Judges wrongdoings, settles disputes, etc.	SP	A
4	Authorizes	Authorizes some behavior (e.g., the building/maintenance of temples, shrines, and other public works; succession and crowning of the king)	SP	A
5	Delegates	Delegates authority/governance	SP	A
6	Controls behavior	Controls the behavior of others (e.g., controls the behavior of royals)	SP	A
7	Coerces	Controls coercive power (police, army, etc.)	SP	A
8	Protects	Protects people and polity, trade routes; ancestors may protect descendants	SP	A
9	Is responsible for	Is responsible for something (e.g., is responsible for maintaining order of the cosmos, people's food supply)	SP	A
10	Patronizes	Acts as a patron by giving gifts, favors, titles, etc.	SP	A
11	Practices polygamy	Marries more than one wife	SP	A
12	Begets	Begets royal children (possible heirs)	SP	A
13	Inherits	Inherits title, property, leadership role, etc.	SP	A/P
14	Is interpreted	Is interpreted by another; does not "speak" for oneself	SP	A/P
15	Receives tribute	Exacts/receives tribute	SP, E	A
16	Exacts tax	Exacts/receives taxes	SP. E	A

~79~

(*continued on next page*)

TABLE 5 (*continued*)

	Role	Definition	Domain	Active/passive
		CORE ROLES (*continued*)		
17	Allies with	Forms alliances with people within/outside territory; via marriage, feasts, hunts, etc. (not gifts)	SP,E	A
18	Controls access	Controls access to critical resources (e.g., water, farmland, or property, especially luxury goods)	SP, E	A
19	Negotiates	Negotiates (e.g., treaties and trade agreements)	SP, E	A
20	Builds/maintains	Builds or maintains buildings, infrastructure including an army, temples, shrines	SP, R	A
21	Owns land/property	Owns productive land, property: slaves, palace, business	E	A
22	Gains wealth	Acquires wealth through king's favor, spoils of war, production, etc.	E	A
23	Maintains/expands boundaries	Conquers new territory; maintains/defends boundaries	W	A
24	Leads	Leads people in some endeavor (e.g., military campaign—any level of leadership)	W	A
25	Fights	Fights in battles, etc.	W	A
26	Captures/is captured	Captures territory, prisoners of war, etc.	W	A/P
27	Monopolizes/interprets	Monopolizes access to ancestors and gods; interprets their will	R	A
28	Participates in ritual	Performs, assists, observes ritual or other events	R	A
29	Serves the gods	"Feeds" the gods through offerings of plants, animals, humans	R	A
30	Serves	Serves, carries out (enforces) superior's wishes; a ruler may serve the gods or ancestors.	R	P
		OTHER POSSIBLE ROLES		
31	Predicts	Divines the future (including outcome of war) via the will of the gods (immediate or long term)	SP, E, W	A/P
32	Supervises	Manages a section or government department; supervises workers	SP	A
33	Marries kin	Marries sister/half-sister to show/intensify his divine status	SP	A

~80~

one means of controlling the decision-making process. Success in war or protecting the populace from famine, drought, and other pestilences increases the ruler's prestige and right to rule. Rulers maintain power over the populace by coercion when they have control of the troops, especially if they maintain their own standing army (e.g., Aztec [M. E. Smith 2003:154] and possibly Late Shang China [Keightley 1983:548, 555]). By controlling trade and access to natural or imported resources (which most did), the ruler regulates the distribution of goods and resources, thereby controlling everyone's opportunity for wealth accumulation.   ~81~

The ruler uses different types of control over royals and nobles. He binds them to his will not only through coercion (we assume) but also through patronage (which makes them obligated to him) and marriage alliances.

A ruler's life is not all control and conspicuous consumption, however. With position comes the responsibility to actually protect the populace from risk—starvation or slavery imposed by a conquering ruler—and the nobility from destroying each other or replacing the ruler. So on the one hand, in the words of Mel Brooks, "It's good to be the king!" But on the other, it is a rather risky position. What if the king cannot deliver on his promises? What if he is captured by a foreign power? The literature provides many examples of kings being deposed, or worse.

Before leaving our discussion of rulers, we show which societies share the most roles for a ruler. For this analysis, we reintroduce the Tewa and Hopi as foils for the premodern states. Figure 2 is a dendrogram produced from a Johnson's hierarchical cluster analysis and checked by the UCINET core-periphery test. It shows that Old Kingdom Egypt and Lower Mesopotamia, Benin, and the Classic Maya share the most roles for a ruler. Next, Protohistoric Hawai'i joins the Classic Maya and Benin, and this triad joins Old Kingdom Egypt and Lower Mesopotamia. Late Shang China joins the large cluster, and the other societies add on to it. The exceptions are the Hopi and Tewa, which do not have rulers and therefore appear separate from the other societies in the dendrogram.

Perhaps the similarity between Old Kingdom Egypt and Lower Mesopotamia relates to geographic proximity and intercommunication, but other primary states such as the Classic Maya and Late Shang China were not connected. Yet they exhibit the same pattern. Therefore we can assert with confidence that the roles/behaviors of the ruler status are independent inventions.

~82~   Primary and secondary states exhibit similar role/behavior patterns even though the primary states cluster at the top of the graph. This means that there is little difference in the roles/behaviors played by the leader of a premodern state that developed independently or that adapted a state model from another polity.

Figure 2 also shows that Protohistoric Hawai'i is situated in the middle of the state societies, suggesting once again that it is a state with the same basic organization of leadership as the other premodern states.

All in all, analysis of the ruler's roles/behaviors does not present an earth-shattering new perspective on premodern states. However, it does corroborate the findings of archaeologists such as Peregrine (2012) and Adam T. Smith (2003:108–109).

FARMERS, HERDERS, AND SERFS

We decided the best way to analyze tillers of the soil was to compare farmers and herders with serfs as the two groups engage in the same work but have different statuses, or legal standing. We define farmers as people who have the right to farm a segment of land through either ownership or membership in a settlement or kin group, i.e., usufruct rights. Herders (and Hawaiian pond fishermen) may own part or all of the animals they raise. One may think of farmers and herders as small business owners usually farming on

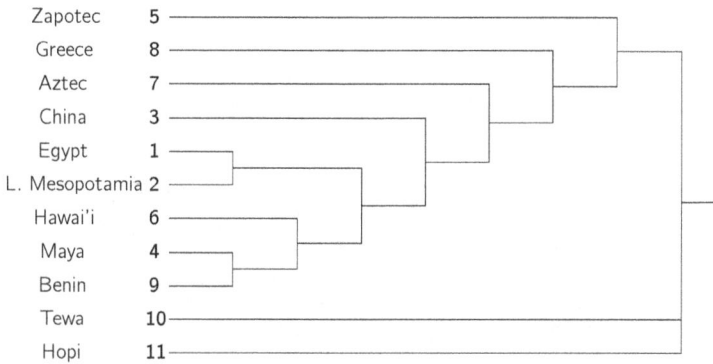

FIGURE 2 Societies whose rulers share roles/behaviors. The dendrogram was constructed from a Johnson's Hierarchical Clustering analysis in UCINET 6 (Borgatti, Everett, and Freeman 2002).

borrowed or leased land or raising their own or borrowed livestock. They are commoners (rank 3) in the sociopolitical hierarchy.

Serfs, on the other hand, till the soil or raise animals owned by an overlord; they are employees. Researchers classify serfs as commoners: Aztec (Berdan et al. 1996:3; M. E. Smith 2003:154); Benin (Bradbury 1973:151–157); Classic Maya (Houston and Inomata 2010:218); Postclassic Zapotec (Whitecotton 1977:149); Old Kingdom Egypt (Baines and Yoffee 1998:229); and Late Shang China (Trigger 2003:157).

Figure 3 represents networks of roles played by farmers/ herders and serfs. We included every role that appears in at least one society in the sample. The nodes in the graphs are the roles and the edges link nodes that co-occur in a particular society. The thicker the edges and the closer the nodes, the more societies include the co-occurring nodes in their database.

Of the nine possible roles/behaviors for farmers and herders, three fall into the sociopolitical domain. All the roles in this domain are passive: "is ruled" by a state government, "is coerced" by the power of an army or police, and "is controlled" by others. This means that when performing their roles, farmers and herders

(a) Co-occurrence of roles/behaviors of farmers/herders

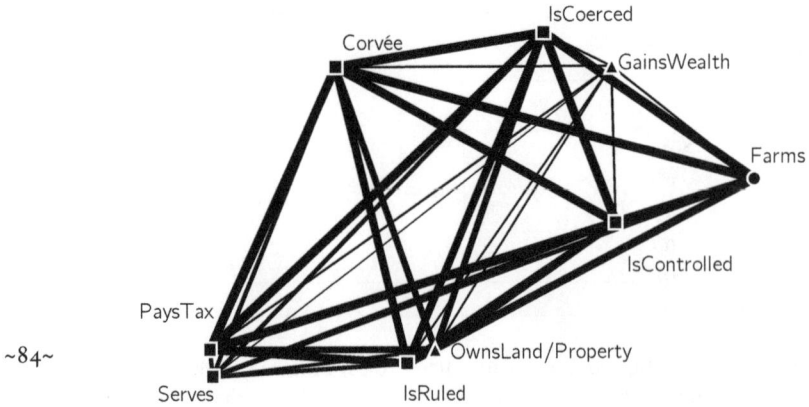

(b) Co-occurrence of roles/behaviors of serfs

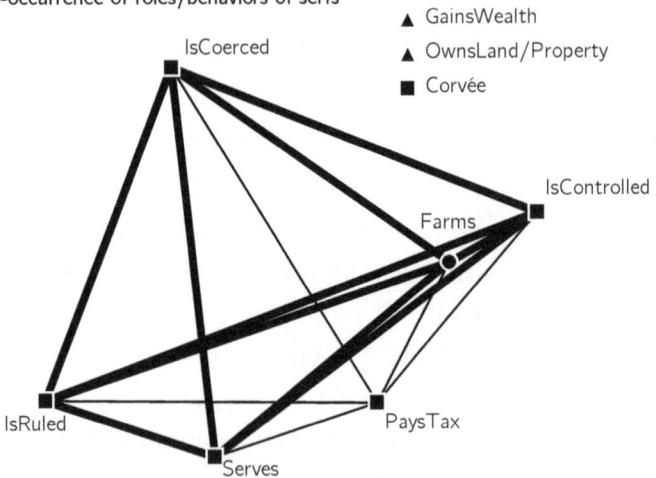

FIGURE 3 Comparison of state-level farmers/herders with serfs by the co-occurrence of the roles/behaviors they exhibited. The nodes are the possible roles/behaviors that farmers/herders (Figure 3a) or serfs (Figure 3b) played. The edges link nodes that co-occur. The thicker the edge and the closer the nodes, the more states include the co-occurring node pairs in the database. Passive nodes are represented by squares; active nodes are triangles; and active roles with a passive component are circles. The networks were generated in NetDraw (Borgatti 2002).

have little or no decision-making authority. These roles form part of the cluster of core farmer/herder roles in Figure 3a.

Two roles, "corvée" and "serves," span the sociopolitical and economic domains. They are partly sociopolitical because people perform the roles as part of their social status. But because the duties associated with the roles are pure labor, they may be considered economic as well. "Corvée" means the labor a farmer/herder owes the state or an overlord each year, while "serves" refers to directly serving a superior. While farmers/herders serve the gods and/or ancestors like anyone in their society, there is no evidence that they perform labor for an overlord while fulfilling their agricultural duties. Of course, while they are performing corvée labor, many serve overlords directly.

~85~

The remaining roles/behaviors fit within the economic domain. Only "gains wealth" is active. It is represented by a triangle in the graph. Although we found no direct evidence of farmers who accumulate wealth, we reasoned that since farmers can become debtors, it follows that they can gain wealth as well. This is the case among the Classic Maya (Houston and Inomata 2010:47), among others. Still, this role is situated far away from the other nodes and has weaker co-occurrence ties than any other node in the graph.

"Farms" may be seen as active or passive. It means to actively make decisions—where and when to plant, water, or fertilize different plant species or how to feed animals (moving them around to different pastures or growing fodder). But farmers and herders inherit their status and have little choice as to whether or not to practice their parents' occupation. "Owns land/property" is also a mixture of active and passive decision-making, for most farmers do not own land in the modern sense of the term, as they cannot buy and sell it. Rather, they have usufruct rights as members of a kin group or settlement. Taking up their usufruct rights may be a choice for some, but most have little mobility because their property rights are tied to their membership in a clan or lineage. (The Classic Maya seem to be an exception [Houston and Inomata 2010:243–244].)

"Pays taxes" is the only economic role that is purely passive, as farmers and herders do not have the right to decide whether or not and how much to pay.

In contrast to the ruler, who has 28 possible active roles/behaviors, two roles/behaviors with active and passive components, and two passive roles/behaviors, farmers and herders have only one active role, two active roles with passive components, and six completely passive roles.

When we compare farmers and herders with serfs, we see that the latter do not have all the roles/behaviors that the first two do. Unlike farmers and herders, serfs lack ownership or usufruct rights to the land they cultivate. Therefore, they neither "own land" nor have the opportunity to "gain wealth." Furthermore, they do not perform "corvée" labor as they are full-time servants. These three roles are off to the side of Figure 3b, for no premodern state serf engages in them. The other five roles/behaviors are all passive because serfs have no choice but to farm or serve their overlords.

Aside from the logical conclusion that serfs neither own land nor perform corvée labor, we have little information on whether or not serfs pay taxes to the state, with the exception of the Postclassic Zapotec, for we have information that serfs do not do so (Whitecotton 1977:150–151). It makes little sense to expect serfs to pay taxes when they are farming someone else's land and have to give all but subsistence fare to their overlord. Serfs play only passive roles. This is interesting because it distinguishes them from farmers/herders. It also differentiates serfs from slaves, who sometimes have active role options (see below).

When we compare roles/behaviors of farmers/herders by premodern state and non-state in Figure 4a, we find that the non-states

(a) Comparison of societies by the roles/behaviors of farmers/herders

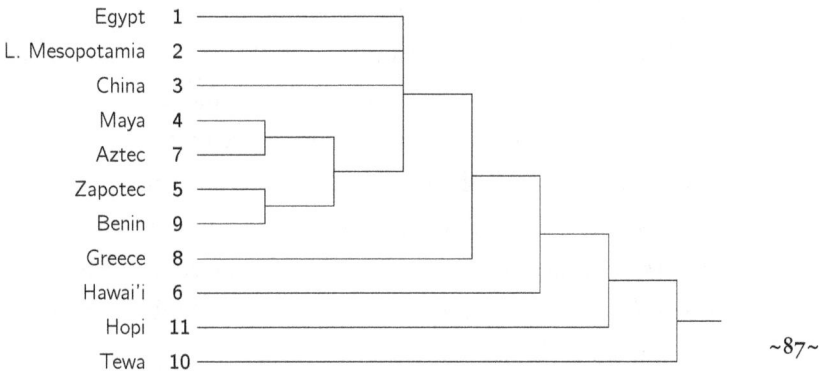

Egypt	1
L. Mesopotamia	2
China	3
Maya	4
Aztec	7
Zapotec	5
Benin	9
Greece	8
Hawai'i	6
Hopi	11
Tewa	10

~87~

(b) Comparison of societies by the roles/behaviors of serfs

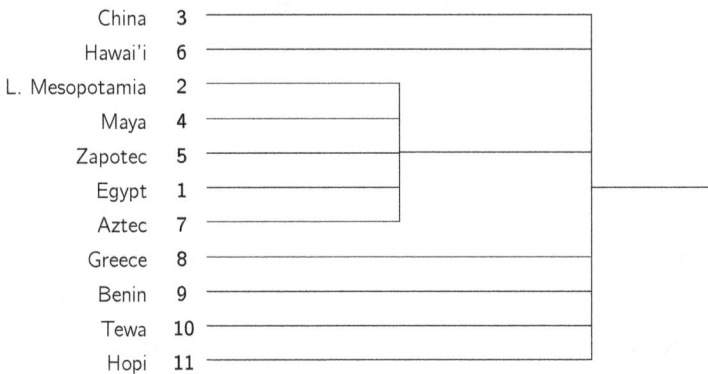

China	3
Hawai'i	6
L. Mesopotamia	2
Maya	4
Zapotec	5
Egypt	1
Aztec	7
Greece	8
Benin	9
Tewa	10
Hopi	11

FIGURE 4 Comparison of farmers/herders (Figure 4a) with serfs (Figure 4b) by society. The more two or more societies share node pairs (co-occurrences), the closer they are to each other in the dendrograms, which were generated in UCINET 6 (Borgatti, Everett, and Freeman 2002).

appear at the bottom of the dendrogram. That is because the non-states do not share with the states the economic roles/behaviors as well as the sociopolitical roles/behaviors found in Figure 3a. The Tewa and Hopi share the roles/behaviors of "farms" and "gains wealth" with the premodern states, but they do not have the roles/

behaviors "is controlled," "is ruled," performs "corvée" labor, or "pays taxes" in the farmer/herder graph.

Figure 4b is the dendrogram for serfs. Here, Late Shang China, Protohistoric Hawai'i, Mycenaean Greece, and Benin join with the Tewa and Hopi, for we have no evidence that any of the state societies have serfs. Perhaps this is the result of semantics used by scholars, but we remained conservative in our interpretation of farmers as commoners or serfs. Looking at the data on Mycenaean Greece (more specifically, Pylos), the Linear B tablets record no mention of serfs. Rather, people "lease" small parcels of land from the damos (community), private landowners, temples, or the palace (Uchitel 2005:474–475). Late Shang China may have day laborers farming the king's land, but there is no report of serfs per se (Trigger 2003:326).

SLAVES AND INDENTURED SLAVES

The definition of slave during the premodern states period is problematic, for some authors conflate slaves and indentured slaves (sometimes called indentured servants) (Pennock 2008:19; Soustelle 1961:74; Aguilar-Moreno 2007:75). While slaves are the property of another and do the work assigned them with little autonomy, indentured slaves or servants are (usually) commoners who sell themselves (or are sold by family members) in order to pay off debts or taxes. During their indenture, they have little or no freedom. But in several societies, they are allowed to work off the debt and regain their freedom. In this section, we distinguish indentured slaves from slaves only in how people obtain their status or are freed from it. The economic roles are the same, for one could argue that an indentured slave is basically performing slave labor during his/her term of servitude.

~89~

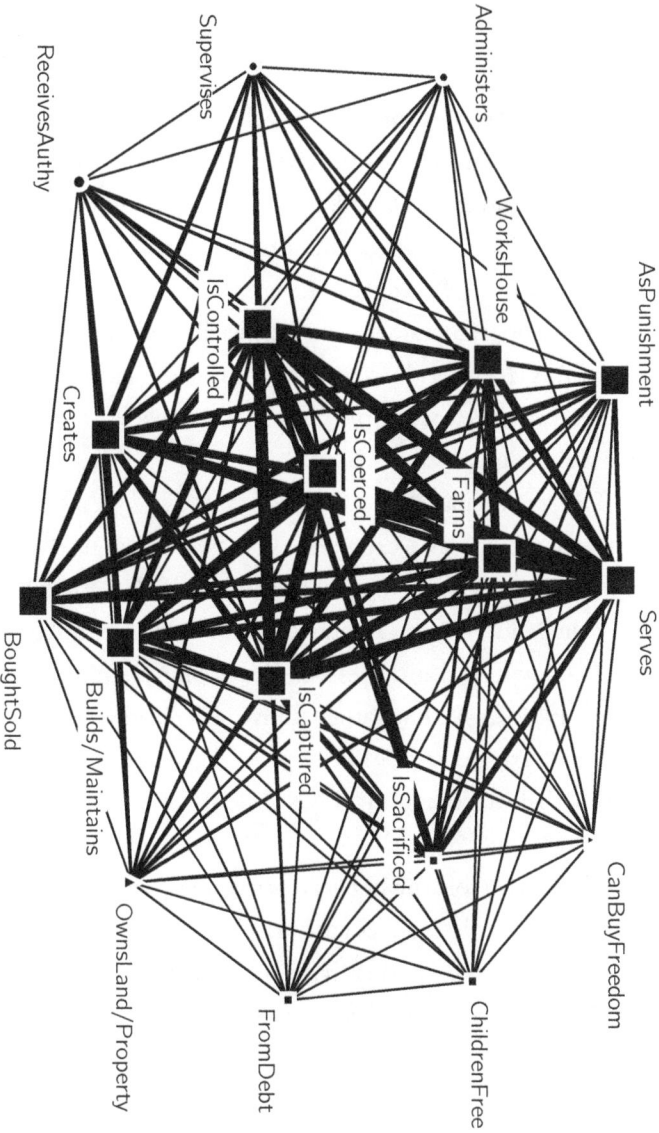

FIGURE 5 Co-occurring roles of slaves and indentured slaves showing core and peripheral roles. Passive nodes are represented by squares; active nodes are triangles; and nodes that may be active or passive are circles. The size of the nodes was determined by a measure of "betweenness." And the thicker the edges, the more societies shared co-occurrences of node pairs. The network was generated in NetDraw (Borgatti 2002).

Slave roles/behaviors of premodern states[2] appear in Figure 5. The square nodes represent passive roles/behaviors. "Is controlled" and "is coerced" refer to the fact that a slave's life is totally controlled by an overlord. The slave roles in the economic domain are largely passive also. These are "serves" a master; "builds/maintains" buildings and infrastructure; works in the house ("WorksHouse"); "farms"; and "creates" pottery, woven cloth, etc. Needless to say, "is sacrificed" is definitely not the choice of the slave (with some exceptions, we would expect). Still, only four of the eight states use slaves as sacrifice: Late Shang China, Protohistoric Hawai'i, Postclassic Zapotec, and the Aztec. Old Kingdom Egypt and Early Dynasty Lower Mesopotamia no longer sacrifice humans and so do not sacrifice slaves. Neither does Mycenaean Greece. Note that all of the large nodes in the core of the network are squares, that is, represent passive roles/behaviors.

The triangular nodes represent active roles/behaviors, or actions in which slaves can make their own decisions. "Owns land/ property" seems to be limited to Mycenaean Greece, where slaves can own land (Deger-Jalkotzy 1972:147), and the Aztec, where slaves can own land and livestock (M. E. Smith 2003:137, 151–152; Hassig 1992:137). The other active role/behavior is "Can buy freedom." Trigger (2003:159) reports that Aztec indentured slaves can buy their freedom, but this is the only case we could find in the literature.

Three slave roles lie between active and passive. "Administers," "supervises," and "receives authority" mean that a slave can direct the work of others, whether in the bureaucracy or the work gang. As mentioned above, Benin slaves could hold top military positions and command armies until the eighteenth century (Osadolor 2001:9). Although the status holder may have significant power and

---

[2] Because the Tewa and Hopi do not have slaves, they are not included in this part of the analysis.

command respect, he/she still answers to someone above (see, for example, Adams 2010: §4.3 on Lower Mesopotamia). Thus the role has both active and passive components.

The degree to which roles/behaviors are active or passive seems in part to be determined by the slaves' position in the hierarchy, which can differ depending on who owns the slave. For example, slaves of the nobility in Benin can have active roles as war commanders or other top military officials (Osadolor 2001:9). Slaves may also have certain skills prior to their enslavement that allow them to occupy certain statuses. Those who know a craft—how to build a boat, chariot, or bronze vessel—may have higher rank than slaves who work in construction or agriculture. In addition, some high-rank people who become enslaved may retain some aspects of that rank while their roles change.

~91~

In the premodern states, slave status is imposed on people rather than chosen. All nine premodern states enslave captives in war or raids. Old Kingdom Egypt, Protohistoric Hawai'i, and the Aztec punish people for crimes or indebtedness by enslaving them. The Aztec (M. E. Smith 2003:137; Aguilar-Moreno 2007:75) and the Classic Maya (Houston and Inomata 2010:47) sell family members or themselves into slavery in order to pay off a debt. Lower Mesopotamia, the Postclassic Zapotec, the Aztec, and Benin buy and sell (trade) slaves.

The premodern states in Figure 5 form a hierarchy of similarity clusters, reflecting the complexity of the slave status and the conflation of "indentured slave" with "slave" here.

Figure 6 shows the Euclidean distance between all the societies of this study. It exhibits the relative similarity of slave traits between Lower Mesopotamia and Benin, with the Aztec joining the two at the next level. These three share two roles that few of the others do, namely "owns property" and "is bought/sold." Mycenaean Greece,

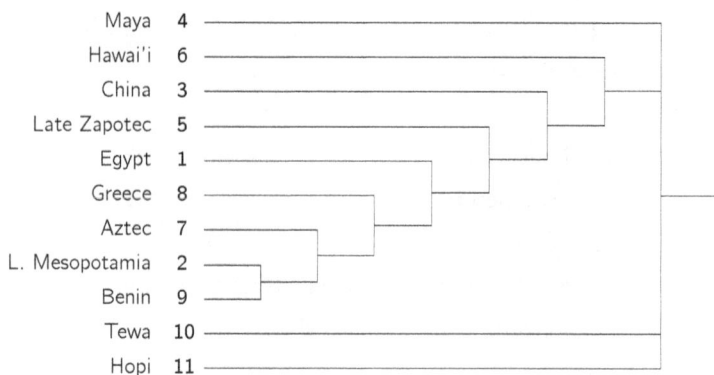

```
Maya        4  ─────────────────────────────────┐
Hawai'i     6  ─────────────────────────────────┤    ┐
China       3  ───────────────────────────────┐ │    │
Late Zapotec 5 ─────────────────────────────┐ ├─┘    │
Egypt       1  ─────────────────────────┐   ├─┘      │
Greece      8  ─────────────────────┐   ├───┘        ├──
Aztec       7  ───────────────────┐ ├───┘            │
L. Mesopotamia 2 ──────┐          ├─┘                │
Benin       9  ────────┴──────────┘                  │
Tewa        10 ──────────────────────────────────────┘
Hopi        11 ──────────────────────────────────────
```

FIGURE 6 Dendrogram of societies that treat the roles/behaviors of slaves and indentured servants in a similar manner. The graph was generated in UCINET 6 (Borgatti, Everett, and Freeman 2002), using Johnson's Hierarchical Cluster analysis.

then Old Kingdom Egypt, Postclassic Zapotec, Late Shang China, and Protohistoric Hawai'i join this core.

The dendrogram reflects what we know of slave (and indentured slave) statuses and roles, but it does not reflect the different societal structures, for we know that Mycenaean Greece and Late Shang China are structurally different from Lower Mesopotamia and Old Kingdom Egypt. What the graph seems to reflect is the states' philosophy of humanity. All the states with slaves seem to see captives in war or raids as fair game—the old "us vs. them" idea. As far as we know, only Lower Mesopotamia, the Aztec, Classic Maya, Benin, and the Postclassic Zapotec trade slaves like chattel, and only the Aztec, Classic Maya, Benin, and Protohistoric Hawai'i enslave debtors or criminals. Although the Aztec use slaves as human sacrifice, they also allow indentured slaves to buy their freedom.

Conclusion

We have undertaken this exercise in the hope that a status, role, and behavior comparison of premodern states according to a modified version of Linton's role theory and the use of network theory would yield new insights into their organization, for we have been able to corroborate some archaeological insights/theories and discover some new ones as well.

In order to make Linton's ideas applicable to early complex society, we investigate status in relation to rank. That is, instead of viewing the statuses in any society as a list, we add a second column for rankings with the understanding that some statuses (such as father or weaver) can be held by people in more than one rank. Linton's classification can be made even more useful if we add "imposed" to his classification of achieved and ascribed statuses (1936:118–21). With these new categories, we capture how status is attained for almost the entire society because captives, slaves, and women whose kin choose their mates do not willingly seek their statuses.

~93~

By extracting the statuses, roles, and behaviors from the historical and archaeological literature and by applying statistical and network analysis to the data, we gain a method that helps us visualize the similarities and differences among the premodern states as well as some non-states. While this is not the first comparative methodology proposed (see, for example, Peregrine, chapter 9 in this volume; chapters in M. E. Smith [2012]; and Blanton and Fargher [2008] for some other approaches), it is a simple way to conduct multistate comparison that allows us to break away from older taxonomic thinking.

The methodology has another advantage, for it yields measurable differences among premodern states and between states and non-states. Through measurement, we can judge which societies belong among the states. For example, we see that Protohistoric

Hawai'i institutionalized most of the same statuses and roles as other premodern states and therefore should be considered one. This supports the findings of the leading archaeologists of the Hawaiian past, Robert Hommon (2013) and Patrick Kirch (2010).

Beyond methodology, we discovered that searching for the roles/behaviors attached to different statuses could yield insight into decision-making. Roles/behaviors appear to be active, passive, or some combination of the two. By this we mean that a person in a particular status has the ability or power to decide whether or not and how to act. Active roles allow the player to choose whether or not to perform them and, if so, how to do so. The higher a person's rank, the more decision-making ability he/she has. A ruler has the most roles/behaviors open to him/her, and most of these roles are active. The exceptions are the few roles he/she plays vis-à-vis the gods. However, these passive roles add to the ruler's legitimacy and give him/her control of decision-making for the polity. Statuses below the upper echelon exhibit fewer active roles. Farming, for example, may shift from partly active to totally passive depending on a person's status and rank. Farmers have some active roles but serfs do not.

Other social scientists have used this active-passive dichotomy to analyze how people behave in certain circumstances (e.g., Richardson 1985:163–179; Adler, Kless, and Adler 1992), or they associate action or dominance with men and passivity or subordination with women (Linton 1936:99–105, 116–118; Ghvamshahidi 1995). The active-passive dimension of roles/behaviors extends all the way through a society's hierarchy. It allows us to add the critical dimension of power to an analysis, and it adds a valuable tool for cross-cultural comparison.

We found that one of the key differences between non-states and premodern states is an expanding differentiation of legal and economic possibilities. First and foremost, types of rights to arable land and pasture expand in complex societies. In non-states, rights are usually determined by one's membership in a kin group. Writing about the Tsembaga Maring of New Guinea, Johnson and Earle (2000:187–188) note that the owner of arable land is the clan, which "defines ownership rights and restricts access to land." But in a state, the ruler, the royal family, nobles, and temples may own some of the land while kin groups own the rest. In Old Kingdom Egypt and Lower Mesopotamia, the ruler theoretically owns all the land and everyone works for him (Blanton and Fargher 2008:147; Henry Wright, personal communication, March 2014; Trigger 2003:334).

~95~

We also found that diversification in employers was correlated with premodern statehood. Whereas people would work for their families and kin groups in non-states, people in states would work for anyone from a kin group to the state, a temple, or an overlord.

In the sociopolitical domain, we noticed the strong connection between warfare and long-term alliances that are anchored by the exchange of women. Indeed, the extensive practice of forming alliances through marriage suggests that they are a necessary part of warfare, for both forestall war and increase a state's chance of military success. We therefore urge archaeologists to analyze warfare in light of alliance building.

Future research may show that the difference between societies with slaves and those with both slaves and indentured slaves reveals their sense of who is human, who is "us" rather than who is "other." By comparing the hierarchy of gods or deities with the social hierarchy and especially the treatment of slaves, we may learn more about different states' political and religious philosophy. ⸙

ACKNOWLEDGMENTS

Thanks to our researchers Robert Weiner, Kong Fai Cheong, and Jonah Nonomaque. Thanks also to our citizen scientists Jeffrey Cohen, George J. Haddad, Jack M. Jackson, and Shelley Waxman. Other contributors to particular cultures include Sandra Barnes, Laurel Bestock, Gary Feinman, Michael Galaty, Abigail Holeman, Patrick Kirch, Peter Peregrine, Gideon Shelach, Adam D. Smith, Michael E. Smith, Charles Stanish, Stephen Tinney, and John Ware. Special thanks to Mirta Galesic and Henrik Olsson for constructing Table 4 and Figure 1 and for reading several drafts of the paper. Henry Wright, Tim Kohler, Doug White, and Jerry Sabloff also provided sound advice on a draft of the paper. Of course this project would not have been possible without the continual guidance of Jerry Sabloff and Henry Wright.

REFERENCES CITED

Adams, Robert McC.
    2010   Slavery and Freedom in the Third Dynasty of Ur: Implications of the Garshana Archives. *Cuneiform Digital Library Journal* 2010:2. http://cdli.ucla.edu/pubs/cdlj/2010/cdlj2010_002.html. Accessed January 29, 2014.

Adler, Patricia, Steven Kless, and Peter Adler
    1992   Socialization to Gender Roles: Popularity among Elementary School Boys and Girls. *Sociology of Education* 65(3):169–187. http://www.jstor.org/stable/2112807. Accessed March 30, 2015.

Aguilar-Moreno, Manuel
    2007   *Handbook to Life in the Aztec World*. Oxford University Press, Oxford.

Allen, Mark, and Elizabeth Arkush
    2006   Introduction: Archaeology and the Study of War. In *The Archaeology of Warfare: Prehistories and Raiding and Conquest*, edited by Elizabeth Arkush and Mark Allen, pp. 1–19. University Press of Florida, Gainesville.

Ames, K. M.
    2008   The Archaeology of Rank. In *Handbook of Archaeological Theories*, edited by R. A. Bentley, H.D.G. Maschner, and C. Chippindale, pp. 487–514. AltaMira Press, Plymouth, UK.

Arkush, Elizabeth N., and Mark W. Allen
    2006   *The Archaeology of Warfare: Prehistories and Raiding and Conquest*. University Press of Florida, Gainesville.

Baines, John, and Norman Yoffee
    1998   Order, Legitimacy, and Wealth in Ancient Egypt and
    Mesopotamia. In *Archaic States*, edited by Gary M. Feinman and Joyce
    Marcus. School of American Research Press, Santa Fe, New Mexico.

Berdan, Frances F., et al.
    1996   *Aztec Imperial Strategies*. Dumbarton Oaks, Washington,
    DC.

Biddle, B. J.
    1986   Recent Development in Role Theory. *Annual Review of
    Sociology* 12:67–92.

Blanton, Richard, and Lane Fargher
    2008   *Collective Action in the Formation of Pre-Modern States*.   ~97~
    Springer Science + Business Media LLC, New York.

Blau, Judith
    1995   When Weak Ties Are Structured. In *Social Roles and
    Social Institutions*, edited by Judith R. Blau and Norman Goodman.
    Transaction Publishers, New Brunswick, New Jersey.

Borgatti, S. P.
    2002   *Netdraw Network Visualization*. Analytic Technologies,
    Lexington, Kentucky.

Borgatti, S. P., M. G. Everett, and L. C. Freeman
    2002   *UCINET for Windows: Software for Social Network
    Analysis*. Analytic Technologies, Harvard, Massachusetts.

Borgatti, Stephen P., Martin G. Everett, and Jeffrey C. Johnson
    2013   *Analyzing Social Networks*. Sage Publications, Los Angeles.

Bradbury, R. E.
    1973   *Benin Studies*, edited by Peter Morton Williams.
    International African Institute. Oxford University Press, Oxford.

Brughmans, Tom
    2013   Thinking through Networks: A Review of Formal Network
    Methods in Archaeology. *Journal of Archaeological Method and
    Theory* 20:623–662.

Campbell, Roderick
    2015   Animal, Human, God: Pathways of Shang Animality and
    Divinity. In *Animals and Inequality in the Ancient World*, edited by
    Benjamin Arbuckle and Sue Ann McCarty, pp. 251–273. University
    Press of Colorado, Boulder. https://www.academia.edu/2234781/_
    Animal_Human_God_Pathways_of_Shang_Animality_and_
    Divinity_. First accessed June 30, 2014.

Campbell, Roderick, Zhipeng Li, Yuling He, and Yuan Jing
2011   Consumption, Exchange and Production at the Great Settlement Shang: Bone-working at Tiesanlu, Anyang. *Antiquity* 85:1279–1297.

Connell, Samuel, and Jay Silverstein
2006   From Laos to Mesoamerica. In *The Archaeology of Warfare: Prehistories of Raiding and Conquest*, edited by Elizabeth Arkush and Mark Allen, pp. 394–433. University Press of Florida, Gainesville.

Deger-Jalkotzy, Sigrid
1972   The Women of PY An 607. *Minos: Revista de Filologia Egea* 13:137–160. http://dialnet.unirioja.es/servlet/articulo?codigo=3161962. Accessed March 2015.

Dickinson, Oliver
1999   Invasion, Migration and the Shaft Graves. *Bulletin of the Institute of Classical Studies* 43(1):97–107.

Feng, Li
2013   *Early China: A Social and Cultural History*. Cambridge University Press, Cambridge.

Flannery, Kent, and Joyce Marcus
2012   *The Creation of Inequality*. Harvard University Press, Cambridge, Massachusetts.

Freidel, David, and Stanley Guenter
2003   Bearers of War and Creation. *Archaeology* (January). http://archive.archaeology.org/online/features/siteq2/. Accessed March 29, 2015.

Freidel, David, and Linda Schele
1997   Maya Royal Women: A Lesson in Precolumbian History. *Gender in Cross-Cultural Perspective* 1997:59–63.

Gelb, I. J.
1965   The Ancient Mesopotamian Ration System. *Journal of Near Eastern Studies* 24(3):230–243.

Ghvamshahidi, Zohreh
1995   The Linkage between Iranian Patriarchy and the Informal Economy in Maintaining Women's Subordinate Roles in Home-based Carpet Production. *Women's Studies International Forum* 18:135–151. http://www.sciencedirect.com/science/article/pii/027753959580050Y. Accessed March 31, 2015.

Goode, William J.
1960   Norm Commitment and Conformity to Role-Status Obligations. *American Journal of Sociology* 66:246–258.

Goodenough, Ward H.
    1969   Rethinking 'Status' and 'Role': Toward a General Model
    of the Cultural Organization of Social Relationships. In *Cognitive
    Anthropology*, edited by S. A. Tyler, pp. 311–330. Holt, Rinehart and
    Winston, New York.

Hanneman, Robert A., and Mark Riddle
    n.d.   Introduction to Social Network Methods. http://fac-
    ulty.ucr.edu/~hanneman/nettext/C10_Centrality.html. Accessed
    March 15, 2015.

Hassig, Ross
    1992   *War and Society in Ancient Mesoamerica*. University of
    California Press, Berkeley.

Hommon, Robert
    2013   *The Ancient Hawaiian State: Origins of a Political Society*.
    Oxford University Press, Oxford.

Houston, Stephen D., and Takeshi Inomata
    2010   *The Classic Maya*. Cambridge University Press, Cambridge.

Johnson, Allen W., and Timothy Earle
    2000   *The Evolution of Human Societies: From Foraging Group to
    Agrarian State*. Stanford University Press, Stanford, California.

Keightley, David N. (editor)
    1983   *The Origins of Chinese Civilization*. University of California
    Press, Berkeley.

Keightley, David N.
    1999   At the Beginning: The Status of Women in Neolithic and
    Shang China. *NAN NÜ* 1(1):1–63.

Kirch, Patrick
    2010   *How Chiefs Became Kings: Divine Kingship and the Rise
    of Archaic States in Ancient Hawai'i*. University of California Press,
    Berkeley.

Kohler, Timothy
    1992   Fieldhouses, Villages, and the Tragedy of the Commons
    in the Early Northern Anasazi Southwest. *American Antiquity*
    57:617–635.

    2013   How the Pueblos Got their Sprachbund. *Journal of
    Archaeological Method and Theory* 20:212–234.

LeBlanc, Stephen
    2003   Prehistory of Warfare. *Archaeology* (May/June):18–25.

2006   Warfare and the Development of Social Complexity: Some Demographic and Environmental Factors. In *The Archaeology of Warfare: Prehistories of Raiding and Conquest*, edited by Elizabeth Arkush and Mark Allen, pp. 437–468. University Press of Florida, Gainesville.

Linton, Ralph
1936   *The Study of Man*. Appleton-Century-Crofts, Inc., New York.

Lopata, Helena Z.
1995   Role Theory. In *Social Roles and Social Institutions*, edited by Judith R. Blau and Norman Goodman. Transaction Publishers, New Brunswick, New Jersey.

Martin, Simon
2008   Wives and Daughters on the Dallas Altar. *Mesoweb*. http://www.mesoweb.com/articles/martin/Wives&Daughters.pdf. Accessed March 30, 2015.

McAnany, Patricia A.
1995   *Living with the Ancestors: Kinship and Kingship in Ancient Maya Society*. University of Texas Press, Austin.

Merton, Robert K.
1957   The Role-Set: Problems in Sociological Theory. *The British Journal of Sociology* 8:106–120.

Osadolor, Osarhieme Benson
2001   The Military System of the Benin Kingdom, c.1440–1897. Doctoral dissertation. University of Hamburg, Germany. http://ediss.sub.uni-hamburg.de/volltexte/2001/544/pdf/Disse.pdf. Accessed September 2014.

Pennock, Caroline Dodds
2008   *Friends of Blood*. Palgrave Macmillan, New York.

Peregrine, Peter
2012   Power and Legitimation: Political Strategies, Typology, and Cultural Evolution. In *The Comparative Archaeology of Societies*, edited by Michael E. Smith, pp. 165–191. Cambridge University Press, Cambridge.

Richardson, James T.
1985   The Active vs. Passive Convert: Paradigm Conflict in Conversion/Recruitment Research. *Journal for the Scientific Study of Religion* 24(2):163–179.

Sailer, Lee Douglas
1978   Structural Equivalence: Meaning and Definition, Computation and Application. *Social Networks* 1:73–90.

Schele, Linda, and David Freidel
1990   *A Forest of Kings: The Untold Story of the Ancient Maya.* William Morrow and Co., Inc., New York.

Smith, Adam Daniel
2010   The Chinese Sexagenary Cycle and the Ritual Origins of the Calendar. In *Calendars and Years II: Astronomy and Time in the Ancient and Medieval World,* edited by John M. Steele. Oxbow Books, Oxford. http://hdl.handle.net/10022/AC:P:9639. Accessed June 17, 2014.   ~101~

2011   The Ernest K. Smith Collection of Shang Divination Inscriptions at Columbia University, and the Evidence for Scribal Training at Anyang. In *Archaeologies of Text: Archaeology, Technology and Ethics,* edited by Matthew T. Rutz. Oxbow Books, Oxford. http://hdl.handle.net/10022/AC:P:12146. Accessed June 17, 2014.

Smith, Adam T.
2003   *The Political Landscape: Constellations of Authority in Early Complex Polities.* University of California Press, Berkeley.

Smith, Michael E.
2003   *The Aztecs.* Blackwell, Oxford.

2008   *Aztec City-State Capitals.* University Press of Florida, Gainesville.

Smith, Michael E. (editor)
2012   *The Comparative Archaeology of Complex Societies.* Cambridge University Press, Cambridge.

Soustelle, Jacques
1961   *Daily Life of the Aztecs.* Stanford University Press, Stanford, California.

Stanish, Charles, and Abigail Levine
2011   War and Pre-Modern State Formation in the Northern Titicaca Basin, Peru. *Proceedings of the National Academy of Sciences (PNAS)* 108(34):13901–13906.

Trigger, Bruce
2003   *Understanding Early Civilizations.* Cambridge University Press, Cambridge.

Turchin, Peter, Thomas Currie, Edward Turner, and Sergey Gavrilets
2013   War, Space, and the Evolution of Old World Complex
Societies. *PNAS Early Edition*. http://www.pnas.org/cgi/
doi/10.1073/pnas.1308825110. Accessed March 2015.

Uchitel, Alexander
2005   Land-Tenure in Mycenaean Greece and the Hittite Empire:
Linear B Land-Surveys from Pylos and Middle Hittite Land-
Donations. *Journal of the Economic and Social History of the Orient*
48(4):473–486.

Westenholz, Aage
2002   The Sumerian City State. In *A Comparative Study of Six
City-state Cultures: An Investigation*, volume 27, edited by Mogens
Herman Hansen: 23–42. Reitzel, Copenhagen.

Westenholz, Joan Goodnick
1990   Review: Toward a New Conceptualization of the Female
Role in Mesopotamian Society. *Journal of the American Oriental
Society* 110(3):510–521.

Whitecotton, Joseph W.
1977   *The Zapotecs: Princes, Priests and Peasants*. University of
Oklahoma Press, Norman.

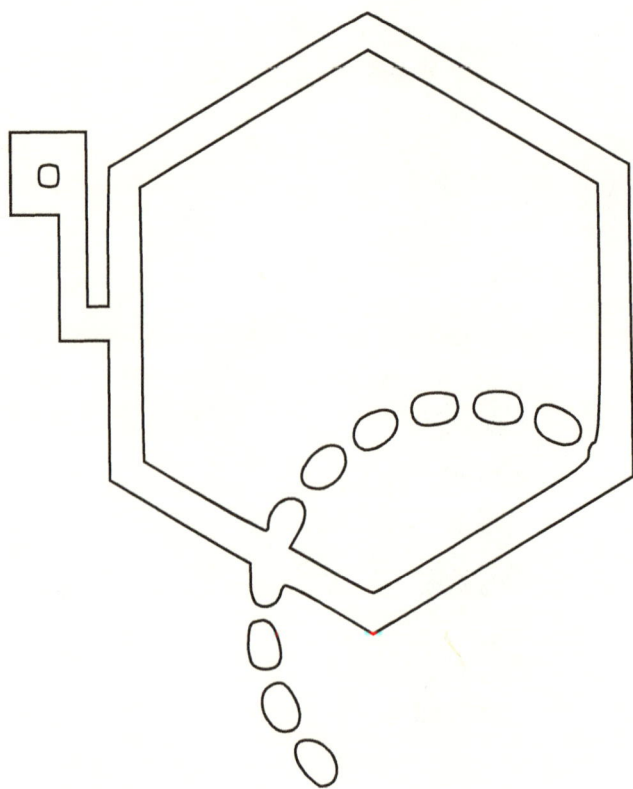

-/

ECOLOGICAL AND SOCIAL DYNAMICS OF TERRITORIALITY AND HIERARCHY FORMATION

Paul L. Hooper, Santa Fe Institute
Eric Alden Smith, University of Washington
Timothy A. Kohler, Washington State University and Santa Fe Institute
Henry T. Wright, University of Michigan and Santa Fe Institute
and Hillard S. Kaplan, Chapman University

The origins of complex societies—those with hierarchically organized political systems, specialized divisions of labor, and relatively
unequal distributions of power and wealth—fascinate us. These
societies tend to be politically and socially integrated at very large
scales, encompassing tens of thousands to hundreds of millions of
individuals. They are the societies in which most of us live.

It hasn't always been this way. For 92 to 98 percent of the existence of modern *Homo sapiens*, it appears that most people lived
in relatively small-scale, egalitarian societies, with muted differences in wealth, status, and political power (Boehm 2001; Smith
et al. 2010). What happened? Why did it happen? These are long-
standing questions, ones that this chapter attempts to address in a
systematic and novel way.

Every state that exists today is a cultural successor to the earliest
states that arose during the Middle Holocene. Each of these states
grew on an economic foundation of intensive agriculture cultivated on well-watered, fertile land (Trigger 2003). Agriculturalists,
however, are not the only societies to manifest political complexity
and inequality.

Despite the fact that most foragers (i.e., hunter-gatherers)
known in recent history live in relatively small-scale, egalitarian
groups, foragers are also capable of generating hierarchical and
unequal societies. Among foragers of the Pacific coast of North

America, multivillage political units encompassing 100 to 10,000 individuals were led by chiefs and exhibited a complex continuum of social rank, slavery, and a significant degree of economic inequality. These foragers relied on resource-producing sites— salmon runs, fruit and nut groves, and coastal sites for maritime hunting and foraging—that were predictably clustered in space and time. This spatial and temporal concentration of resources motivated territorial claims and hierarchically organized, large-scale warfare over access to prime sites (Gunther 1972; Ames 1994; Matson and Coupland 1994: Rick et al. 2005; Kennett 2005).

The common denominator of political complexity among agriculturalists and foragers—the value of claiming and defending durable, defensible resources and resource-producing sites— underlies the central thesis of this chapter: *the social dynamics that arise in the context of economically defensible resources are key generators of large-scale political integration and political hierarchy.*

We establish the theoretical validity of this thesis with a computational model based on a set of simple assumptions. The model identifies features of the natural, social, and technological environment that favor (or disfavor) the development of an arms race between larger and more hierarchically organized territorial coalitions. The following three sections describe the model, summarize its results, then discuss the implications for understanding hierarchy and state formation in human history.

The Model

OVERVIEW

This model represents the cultural evolution of different strategies for gaining access to resource sites. These strategies vary in whether or not they attempt to (1) territorially defend sites, (2) form territorial alliances, and (3) establish territorial political hierarchies (Figure 1 and Table 1). The exogenous parameters of the

ALL STRATEGIES

```
                    ┌──────────────┬──────────────┐
              Nonterritorial    Territorial
                 (T = 0)      (T = 1, μ^T = ?)
                              ┌──────────┬──────────┐
                           Solitary   Alliance-forming
                           (A = 0)   (A = 1, M^max = ?)
```

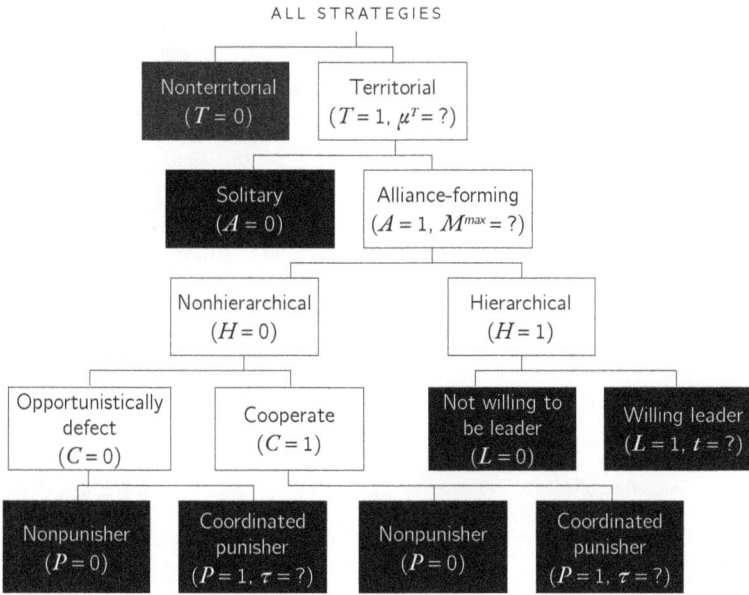

FIGURE 1 A nested taxonomy of strategy types. The black (terminal) nodes indicate fully specified strategy types. The subscript i can be added to any of these variables to indicate the value of individual i. The question marks (?) indicate continuous values that vary across individuals.

TABLE 1 Evolving strategies and decision variables for individual i.

Strategy	Symbol	Values	Associated decision variable	Symbol	Values
Territorial	T_i	0, 1	Contest threshold	μ_i^T	0–1
Alliance-forming	A_i	0, 1	Maximum alliance size	$M_i^{\max}$	0–N
Cooperate	C_i	0, 1	–	–	–
Coordinated punisher	P_i	0, 1	CP coordination threshold	τ_i	0–N
Hierarchical	H_i	0, 1	–	–	–
Willing leader	L_i	0, 1	Tax offered as leader	t_i	0–∞

THE EMERGENCE OF PREMODERN STATES

TABLE 2. Exogenous socioecological parameters.

	Parameter	Symbol	Values
Focal	Variance in patch productivity	$\text{var}(\mu_k)$	0.005, 0.030, 0.055
	Decisiveness of alliances	α	1, 2, 4
	Base cost of alliance*	cA	0.05, 0.15, 0.25
	Cost of cooperation*	cC	0.05, 0.35, 0.65
	Cost of enforcement*	cE	0.05, 0.35, 0.65
	Efficiency of leadership selection process	d	0, 0.5, 1
Additional	Landscape dimensions (in patches)	$x \times x$	33 × 33
	Mean patch productivity	$\text{mean}(\mu_k)$	0.1
	Clustering of productive patches	S	0.1, 0.2, 1.0
	Spatial reach†	r	0.12, 0.36, 0.6
	Intrinsic rate of increase coefficient	B	4
	Rounds per generation	–	25
	Generations per run	–	250
	Probability of mutation	–	0.01

*In units of μ; †In units of x.

TABLE 3. Other variables in the model.

Variable	Symbol	Values
Total number of agents	N	$0-\infty$
Number of agents on patch k	N_k	$0-N$
Number of allies of i	M_i	$0-N$
Number of active defectors in i's alliance	M_i^D	$0-N$
Number of willing punishers in i's alliance	M_i^P	$0-N$
There are ≥ 1 willing punishers in i's alliance	$\hat{P}(1)$	0, 1
There are ≥ τ_i willing punishers in i's alliance	$\hat{P}(\tau_i)$	0, 1
Number of followers of i	F_i	$0-N$

model are intended to reflect variation in the social and ecological circumstances faced by human groups according to time and place (Table 2). The economic defensibility of sites is represented in terms of the variance in the expected productivity of resource patches. Whether or not agents form alliances to claim and defend territory depends on this variance in the quality of land (or "patchiness"), as well as the costs of maintaining and cooperating within alliances. Whether or not individuals are willing to buy into systems of political hierarchy that monitor and enforce the cooperation of alliance members depends on the relative efficiency and costs of hierarchy compared to nonhierarchical alternatives.

~109~

The current model synthesizes the theories of economic defensibility and animal contests (Brown 1964; Maynard Smith and Price 1973; Dyson-Hudson and Smith 1978; Boone 1992) with models of the evolution of cooperation, punishment, and political hierarchy from evolutionary anthropology (Smith and Choi 2007; Hooper et al. 2010; Boyd et al. 2010). The model nests the strategies and replicator dynamics of evolutionary game theory (McElreath and Boyd 2007; Gintis 2009) within a simulation in NetLogo 5.1.0 with explicit spatial and demographic dynamics (Wilensky 1999). This model builds on the base of previous evolutionary models by representing explicit landscapes with productivity that varies in space and time; endogenous returns to scale in competition for territorial resources; and a statistical analysis of stochastic outcomes as a function of socioecological parameters. Like all useful models, it aims to establish clear principles that sharpen our understanding of real-world phenomena, rather than recreate the full complexity of reality itself.

SOCIOECOLOGY

The socioecology of the model is defined by the spatial variability of resource production and an array of parameters affecting the benefits and costs of different social behaviors (Table 2). The exogenous nature of these parameters allows a principled statistical analysis of the results and avoids issues of inference under endogeneity.

The spatial landscape of the model is defined as an $x \times x$ torus containing x^2 patches. In each round, each patch k produces one resource unit (productivity $L_k = 1$) with probability μ_k and produces nothing ($L_k = 0$) with probability $1 - \mu_k$. The spatial distribution of mean productivity μ_k (equal to the probability of producing a resource in each round) is taken as an exogenous characteristic of ecology. The present analysis employs a set of nine landscapes,

specified in terms of the spatial distribution of μ_k (Figure 2). These landscapes have the same overall mean productivity of patches mean(μ_k) but different values for both the "patchiness" of the landscape (i.e., variance in productivity across patches: var(μ_k)) and the extent to which productive patches cluster together in space (S). To generate these landscapes, the product of two orthogonal sine waves (both with the same frequency tuned by the parameter S) was raised to a power Φ ($0 < \Phi < 20$), then renormalized to achieve the target values of mean(μ_k) and var(μ_k). While the theory of economic defensibility emphasizes variance in resource production across both space and time, for simplicity, the current model focuses only on spatial variance; its main results, however, can easily be reinterpreted in terms of temporal variability (or predictability), with some important differences discussed in Dyson-Hudson and Smith (1978).

STRATEGIES AND INTERACTION

The landscape is seeded with a small number of nonterritorial agents, interpretable as single individuals or family units. In each nonoverlapping generation, agents interact over multiple rounds (default = 25). In each round, in random order, each agent attempts to move to an undefended patch, claim a defended patch, or stay in place, depending on its inherited strategy types and decision variables (Figure 1 and Table 1).

Each agent i employs either a nonterritorial strategy ($T_i = 0$) or a territorial strategy ($T_i = 1$). Nonterritorial agents attempt to move to (or stay on) an undefended patch within their spatial reach (r) that yields the highest per capita productivity (L_k/N_k, where N_k is the number of agents occupying patch k). Nonterritorial agents are thus facultatively nomadic, moving when productivity can be improved and staying when local productivity remains high.

~III~

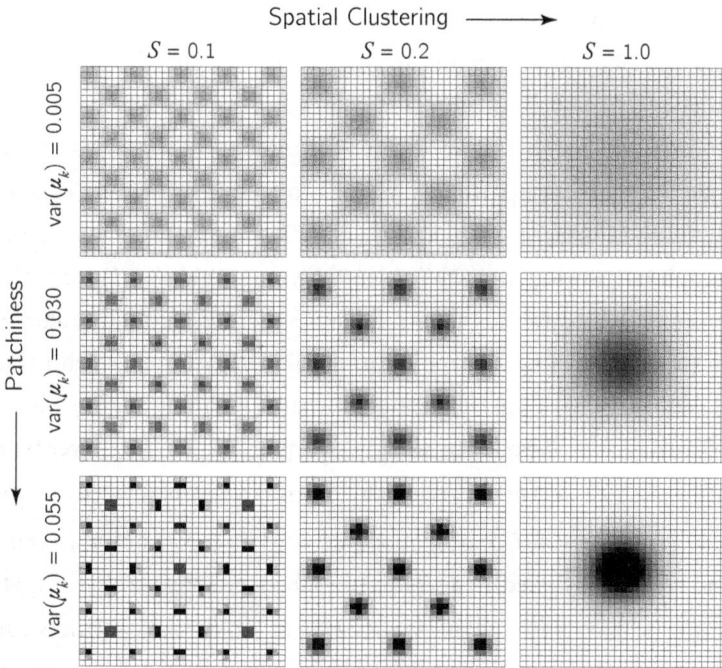

FIGURE 2 Nine ecologies differing in patchiness, var(μ_k), and spatial clustering of productive patches, S. Patchiness increases from top to bottom, while clustering increases from left to right. The mean productivity of each landscape, mean(μ_k), is held constant at 0.1.

Territorial agents that have not claimed a patch first attempt to find an undefended patch within their spatial reach that has expected productivity greater than or equal to their heritable contest threshold (i.e., $\mu_k \geq \mu_i^T$). If an acceptable undefended patch is available, they occupy and claim the patch. If no acceptable patch is found, they attempt to find an already-defended patch within their spatial reach for which $\mu_k \geq \mu_i^T$. If an acceptable patch is found, they contest the existing claimant for ownership. The strength of an agent in contesting a patch depends on whether it employs a solitary strategy ($A_i = 0$) or an alliance-forming strategy ($A_i = 1$). The strength of solitary agents contesting patches is equal to 1. The strength of alliance-forming agents is equal to 1 plus the number of alliance partners that cooperate in contesting the patch.

Prior to a contest, an alliance-forming agent i that has fewer than $M_i^{\max}$ alliance partners attempts to recruit additional alliance partners. Nonhierarchical agents ($H_i = 0$) form alliances with other nonhierarchical agents, while hierarchical agents ($H_i = 1$) form alliances with hierarchical agents. Alliance partnerships (i.e., edges in an alliance network or graph) require pairwise stability as defined in Jackson (2008). Thus, if there is an alliance-forming agent j within i's spatial reach who is not already claiming a patch, has fewer than $M_i^{\max}$ alliance partners , and is able to find a patch adjacent to the patch targeted by i that is above its contest threshold μ_j^T, then i and j become linked in an offensive alliance. This process repeats until i attains $M_i^{\max}$ alliance partners , or no further partners can be found. Once i and their M_i alliance partners have targeted a set of defended patches, each defender p that forms alliances and has fewer than $M_p^{\max}$ partners attempts to recruit additional partners. New defensive alliance partners are required to occupy patches adjacent to p or another member of p's alliance.

Once alliances have formed, a series of $(1 + M_i)$ contests are played out between the offensive and defensive alliances, one for each patch targeted by the members of i's alliance. The strength of i's alliance σ_i is the sum of i's own effort and the effort of those alliance members who are cooperative either by heritable strategy (i.e., $C_i = 1$) or because their cooperation is enforced by coordinated punishers or a hierarchical leader. If M_i^D is the number of active defectors (opportunistic defectors whose cooperation is not enforced), then the strength of the alliance is $\sigma_i = 1 + M_i - M_i^D$.

Cooperation within nonhierarchical alliances can be enforced by the action of coordinated punishers (for whom $P_i = 1$). As in Boyd et al. (2010), a coordinated punisher is willing to expend the effort to enforce cooperation when there are at least τ coordinated punishers present in the alliance, where τ is a heritable decision variable held by the coordinated punisher. The cost of enforcing the cooperation of each of M^D opportunistic defectors, $c^E M^D$, is divided evenly among the willing coordinated punishers in the alliance (i.e., $c^E M^D / M^P$ per punisher).

Cooperation within hierarchical alliances can be enforced through the action of a leader. A leader is selected among those members of the hierarchical alliance who are willing to lead (for whom $L_i = 1$). Each potential leader offers a heritable per capita tax (t_j, defined as a fixed amount rather than a fraction or rate), which is observable to alliance members.

Which potential leader is selected depends on the parameter d reflecting the efficiency of the process that selects leaders within groups. When this process is maximally efficient ($d = 1$), the leader who offers the lowest tax is selected; when $d = 0$, the leader who offers the highest tax is selected; intermediate values of d select the leader closest to that quantile value (e.g., $d = 0.5$ selects the leader offering the median tax). The parameter d can be interpreted as the extent to which alliance members are able to select the most efficient or generous leaders through a low-cost, democratic process (or alternatively as the accuracy of individual's perceptions of the true tax rates). The leader receives a tax t_j paid by each of the F_i other group members (followers) and pays a cost $c^E M^D$ to enforce the cooperation of M^D opportunistic defectors in the alliance. If no members of a hierarchical alliance are willing to lead, the alliance operates as a nonhierarchical alliance, in which cooperation either remains unenforced or is enforced through coordinated punishment.

The probability that i is successful in a contest against p is decided by the contest-success function $\sigma_i^\alpha/(\sigma_i^\alpha + \sigma_p^\alpha)$ (Hirshleifer 2001). The exponent α in this expression represents the decisiveness of alliances in determining the outcome of contests. When α is greater, the slope of the content success function around the inflection point (where the two alliances are of equal strength) is steeper, reflecting more of a winner-take-all environment. α thus reflects the importance of the relative size of the alliance for determining the outcome of the conflict. If the offensive agent is successful, it takes exclusive claim of and occupies the patch, while the defensive agent is ejected to a randomly chosen undefended patch. If the defensive agent is successful, neither move. Members of an offensive alliance that successfully claim new patches remain linked as allies once they have settled on their new territories. Members of defensive alliances that are not ejected from the patches they claim also remain linked as allies following the contest.

At the end of each round, the resources produced by each patch are divided equally among its occupants and added to their cumulative lifetime fitness. Successful territorial agents enjoy the benefit of not having to share with others but must pay the costs of having to claim and defend the land. The per-round contribution to fitness of individual i occupying patch k (equation 1) is thus equal to the per capita productivity of patch k (first term: L_k/N_k) minus the costs of their strategy- and context-dependent behavior: the costs of maintaining alliances (second term: $c^A M_i$);

the cost of cooperating to defend territory, whether dictated by strategy or stimulated by enforcement (third term: $c^C \max(C_i, H_i, \hat{P}(1))$); the coordinated punisher's cost of enforcing the cooperation of each opportunistic defector (fourth term: $c^E \hat{P}(\tau_i) M_i^D / M_i^P$); and the hierarchical agent's cost of being taxed (fifth term: $t_j H^i$). Hierarchical agents who enforce cooperation as leaders also receive the benefit of taxation minus the cost of enforcement for each of F_i followers (sixth term: $(t_i - c^E M_i^D / M_i) F_i$). The individual's per-round contribution to fitness w_i is defined by equation (1):

~115~

$$w_i = L_k N_k - c^A M_i - c^C \max\left(C_i, H_i, \hat{P}(1)\right)$$
$$- \hat{P}(\tau_t) c^E M_i^D / M_i^P - t_j H_t + (t_i - c^E M_i^D / M_i) F_i \qquad (1)$$

REPRODUCTION

After a generation has interacted through multiple rounds, each agent's fitness W_i is calculated as the mean of their contributions to fitness across rounds multiplied by B, a coefficient affecting the intrinsic rate of increase: $W_i = B \times \text{mean}(w_i)$. Each agent produces a whole number of offspring equal to the integer part of W_i, and produces an additional offspring with probability equal to the fractional part of W_i. Offspring are distributed randomly on the landscape.

Offspring inherit their parent's strategies and decision variables, with an independent probability of mutation (default = 0.01) for each variable. Mutations of strategy variables switch between 0 and 1; mutations of the contest threshold μ_i^T and tax offered as leader t_i increase or decrease by 0.01 with equal probability; and mutations of the maximum alliance size $M_i^{\max}$ and the coordinated punisher's coordination threshold τ_i increase or decrease by 1. The older generation dies, territorial claims and alliances are reset, and the new generation begins in round 1.

SAMPLING AND ANALYSIS

The analyses of the simulation in the following section characterize the effects of different socioecological parameters on the principal outcomes of the model, as they evolve through time (Figure 3) and after a long period of evolution (Figures 4 and 5; see also Appendix, p. 306).

To produce the historical trajectories plotted in Figure 3, runs of 250 generations playing 25 rounds per germination were simulated in each of four contrasting socioecologies. Ecology (A) has low patchiness, as well as as low costs of alliance formation, cooperation, and hierarchy. Ecologies (B), (C), and (D) have very patchy landscapes, but vary in terms of the costs of social interaction. Ecology (B) has high costs of cooperation and enforcement and inefficient political institutions; ecology (C) has low costs of cooperation, but high costs of enforcement and inefficient political institutions; while ecology (D) has low costs of cooperation and enforcement and efficient political institutions.

For the statistical analysis of long-run outcomes summarized in Figures 4 and 5, 200 runs of 250 generations playing 25 rounds per generation were parameterized by randomly sampling the parameter values given in Table 2 (see also Appendix, p. 306). For each parameter (with the exception of landscape patchiness) one of three values was sampled with equal probability. For the landscape patchiness, var(μ_k), the high-variance condition was sampled at twice the rate of the medium- and low-variance conditions, to better explore subtleties in the conditions underlying the formation of competitive alliances and hierarchies. Long-run strategy frequencies, mean alliance sizes, and inequality were estimated as a function of socioecological parameters using regression. Inequality was measured as the Gini coefficient of fitness, Gini(W), using the redist package (Handcock 2015). For the analyses of strategy frequencies and alliance sizes in table in the Appendix, p. 309, the lmer function in the lme4 package (Bates et al. 2013) was used to estimate outcomes

in the last 100 generations of each run, including a random effect
to capture the nonindependence of outcomes across generations
within each run

Results

HISTORICAL DYNAMICS

The effects of ecological and social parameters on the population
dynamics of nonterritorial and territorial agents are illustrated
across panels A, B, C, and D in Figure 3. Environment A is char-   ~117~
acterized by a relatively homogeneous distribution of productivity
across the landscape (the landscape in the first row, middle column
of Figure 2). In this environment, nonterritorial agents retain the
advantage over territorial strategies, maintaining a long-run mean
of around two-thirds of the population.

Environment B is marked by a highly patchy distribution of
resources across the landscape (the landscape in the third row,
middle column of Figure 2). Despite the initial nonterritorial
starting point, mutant territorial strategies quickly come to dominate the population. Because environment B has relatively high
costs of cooperation, enforcement, and hierarchy, solitary territorial strategies win out over alliance-forming strategies. Environment
C has the same patchiness as B, but with relatively lower costs of
cooperation and high costs of enforcement and hierarchy. In this
context, alliance-forming territorial strategies outcompete solitary
territorial strategies. Because hierarchy is relatively costly, however,
non-hierarchical agents still outcompete hierarchical strategies.

Environment D has the same patchy landscape as B and C, but
with relatively lower costs of cooperation and enforcement, and
more efficient selection of leaders. After the initial establishment
of territorial and alliance-forming strategies, hierarchical territorial agents come to dominate the landscape. Other territorial and

FIGURE 3 What favors the emergence of hierarchy? For all runs, $S = 0.2$, $\alpha = 4$, and $r = 0.36$. Lines were smoothed by local polynomial regression with loess smoothing parameter 0.3 (R Development Core Team 2008).

~118~

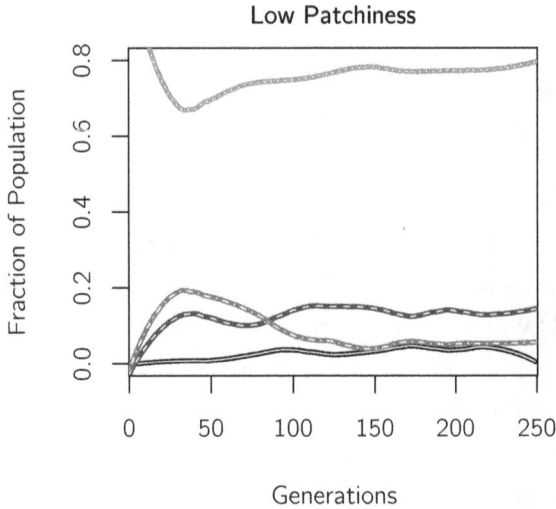

Low Patchiness

(A) An environment with low patchiness (i.e. low variance in the productivity of land, var$(\mu_k) = 0.005$) yet low social costs of alliances and hierarchy ($c^A = 0.05$, $c^C = 0.05$, $c^E = 0.05$, $d = 1$). Nonterritorial agents outcompete territorial strategies and maintain predominance in the population.

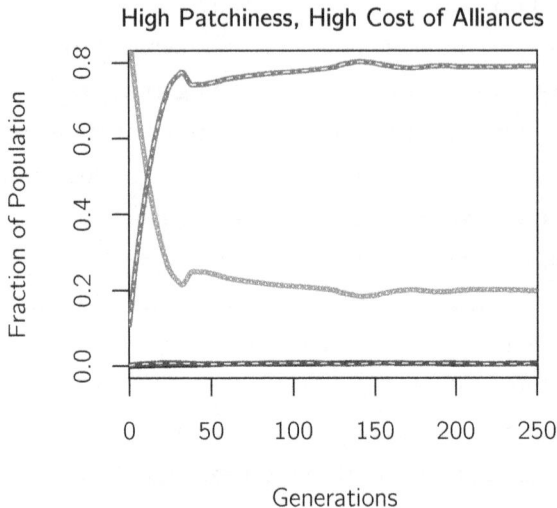

High Patchiness, High Cost of Alliances

(B) An environment with high patchiness (var$(\mu_k) = 0.055$) and relatively high social costs of alliances and hierarchy ($c^A = 0.25$, $c^C = 0.35$, $c^E = 0.35$, $d = 1$). Solitary territorial agents quickly establish a majority in the population.

Nonterritorial

Solitary Territorial

Nonhierarchical Allied

Hierarchical

High Patchiness, High Cost of Hierarchy

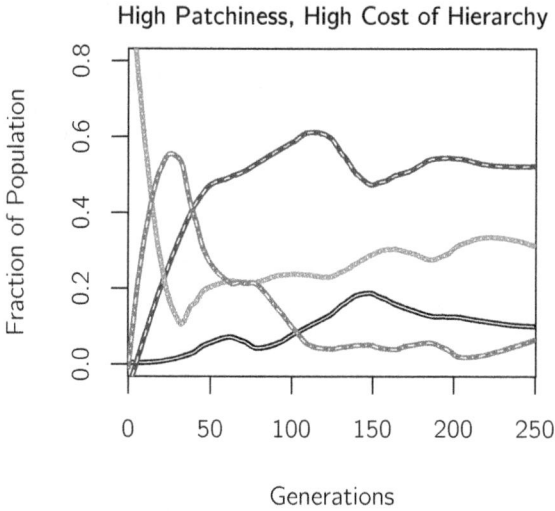

Generations

(C) An environment with high patchiness (var(μ_k) = 0.055), low costs of alliances (c^A = 0.05, c^C = 0.05), and high costs of hierarchy (c^E = 0.35, d = 1). Following an initial invasion of solitary territorial agents, non-hierarchical alliance-forming territorial agents form the majority of the population over time.

High Patchiness, Lower Cost of Hierarchy

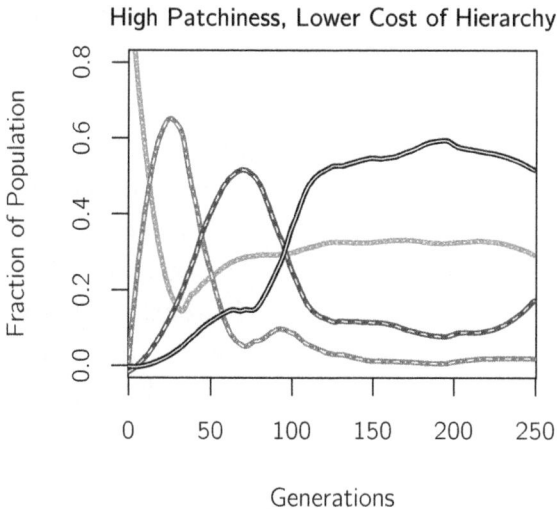

Generations

(D) An environment with high patchiness (var(μ_k) = 0.055) and relatively low costs of alliances and hierarchy (c^A = 0.05, c^C = 0.05, c^E = 0.05, d = 1). After initial invasions by solitary and alliance-forming territorial agents, hierarchical strategies dominate the population.

nonterritorial strategies are maintained in the minority through frequency-dependent selection.

The relationship between alliance size and strategy frequencies across ecologies can be summarized in three points. First, there is a general tendency for the size of alliances between territorial agents to first increase through time, then level off in dynamic equilibrium. Second, the long-run size of alliances varies across ecologies, with smaller groups in ecologies A and B, larger groups in ecology C, and the largest groups in ecology D. Third, hierarchical strategies consistently maintain larger alliances than nonhierarchical alliance-forming strategies. Hierarchy evolves in the context of and reinforces large territorial alliances

LONG-RUN OUTCOMES

The effects of patchiness and social costs of hierarchy on the long-run frequency of different strategies is illustrated in Figure 4. Territoriality always increases (while nonterritoriality decreases) with greater patchiness. Whether nonhierarchical or hierarchical territorial strategies are favored depends crucially on the costs of alliances and hierarchy. Hierarchical territorial strategies become common only when the costs of alliances and hierarchy are not prohibitively high.

The statistical analysis of long-run outcomes indicates that the effects of social parameters (the costs of alliances, cooperation, and enforcement and the decisiveness of alliances) on sociopolitical outcomes depend crucially on the underlying degree of patchiness (see Appendix, p. 306). In landscapes with greater patchiness, there are negative effects of the costs of alliance formation and the costs of cooperation on the frequency of territorial, alliance-forming, and hierarchical strategies. These effects are absent or muted in landscapes with low patchiness.

In patchier environments, the decisiveness of alliances drives

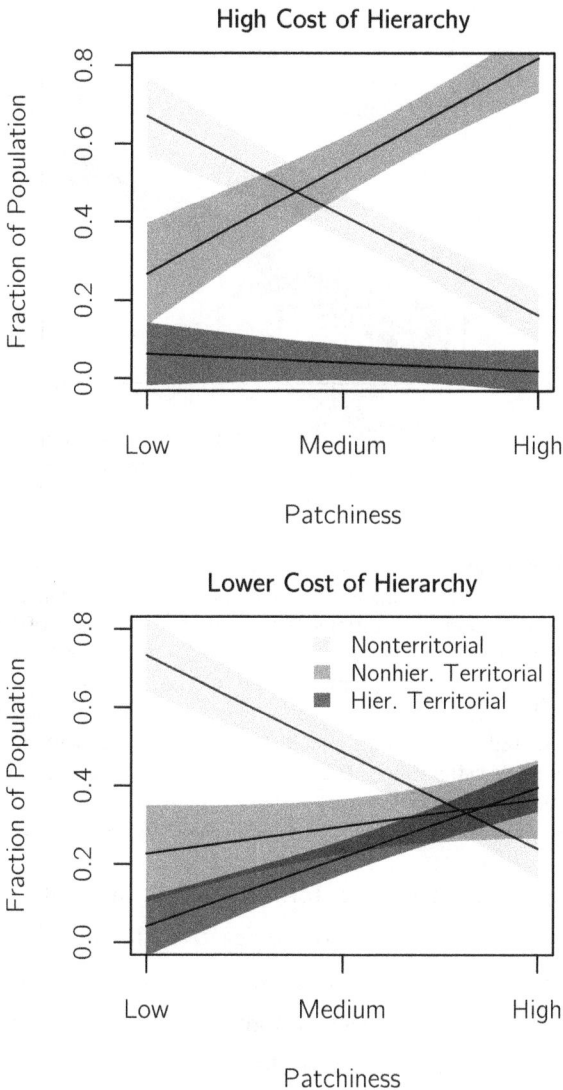

FIGURE 4 How do patchiness and the cost of hierarchy affect the evolution of territorial and hierarchical strategies? In both panels, increased patchiness results in fewer nonterritorial and more territorial strategies in the long run. Panel (A): When alliances and hierarchy are more costly ($c^A = 0.35$, $c^C = 0.35$, $c^E = 0.35$, $d = 0$), hierarchical strategies are always outcompeted by nonhierarchical strategies. Panel (B): When alliances and hierarchy are relatively less costly ($c^A = 0.05$, $c^C = 0.05$, $c^E = 0.05$, $d = 1$), hierarchical strategies become common at higher levels of patchiness. In both panels, $S = 0.2$, $\alpha = 4$, and $r = 0.36$. Lines and shading indicate predicted mean values and 95% confidence intervals from linear regression models estimating strategy frequencies after 250 generations.

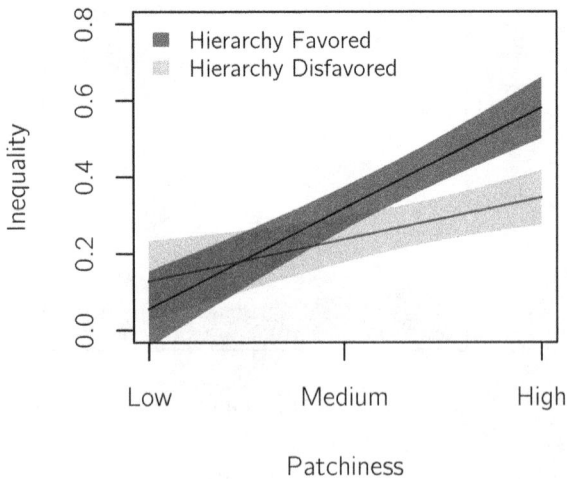

FIGURE 5 How do patchiness and hierarchy affect inequality? Inequality increases with patchiness in all environments. Patchiness leads to even greater inequality under conditions that favor the evolution of hierarchy (dark gray: $c^A = 0.05$, $c^C = 0.05$, $c^E = 0.05$, $d = 1$) compared to conditions that do not favor hierarchy (light gray: $c^A = 0.35$, $c^C = 0.35$, $c^E = 0.35$, $d = 0$). For both lines, $S = 0.2$, $\alpha = 4$, and $r = 0.36$. Lines and shading indicate predicted mean values and 95% confidence intervals from linear regression models estimating the Gini coefficient of fitness, Gini(W), after 250 generations.

higher frequencies of alliance–forming and hierarchical strategies. In very patchy landscapes, there are negative effects of the costs of enforcement on alliance-forming and hierarchical strategies, and positive effects of the efficiency of political processes in hierarchical groups on the frequency of hierarchical preferences.

The statistical analysis (see Appendix, p. 306) confirms the previous insight that the long-run size of alliances increases with landscape patchiness. Social and ecological parameters interact to determine alliance size: in patchier environments, alliance size increases with greater decisiveness of alliances, and decreases with the costs of alliance formation, cooperation, and enforcement. The results confirm that alliances grow to larger scales when the benefits (in terms of access to land) are high, and the social costs of alliance formation are not prohibitive. At very large scales, hierarchical alliances outcompete nonhierarchical alliances due to their efficiency in organizing for collective action in territorial competition.

Inequality is inextricably linked to hierarchy across the worlds simulated in this model, as shown in Figure 5. While hierarchical institutions grow in the context of high patchiness and inequality, they also amplify inequalities. In this simulation, hierarchy increases the Gini by roughly 50 percent, on top of the direct effects of patchiness and territoriality. This would not be the case if members of hierarchies could always choose the most efficient leaders at zero cost (Hooper et al. 2010).

Conclusions

The ecological and social dynamics that drive this model may be fundamental to the history of social hierarchy in our species. The model shows that hierarchy and inequality develop regularly in ecologies that favor the formation of coalitions to defend and contest resources.

Hierarchies for competition over resources are particularly likely where egalitarian means of promoting cooperation in coalitions are ineffectual and hierarchies are not unbearably costly. Hierarchies become more costly when group members are restricted in their ability to choose efficient leaders. In the real world, as in the model, hierarchical institutions that are inefficient relative to other feasible social arrangements often face overthrow, reform, or extinction. The difficulty and cost of these transitions, however, allows persistence of the kind of oppressive hierarchies observed countless times throughout history.

The model illustrates the interaction of ecological and social dynamics in two senses. First, the analysis shows the statistical interaction between ecology—in terms of the variance of land productivity—and other socioecological parameters, such as the cost of social interactions and the decisiveness of contests (see table in Appendix, p. 306). Second, the endogenous social dynamics are the very mechanism by which the effects of the exogenous ecological parameters are manifest through time (Figure 3); conversely, the effects of the social dynamics necessarily depend on the presence of favorable socioecological conditions. Thus, neither natural ecology nor social behavior alone is sufficient to explain the central result of the model: the emergence of norms and institutions for collective action, including hierarchy, in the presence of concentrated and predictable resources.

~124~

We have employed a definition of socioecology as the relationship between an organism and its natural, constructed, and social

---

At very large scales, hierarchical alliances outcompete nonhierarchical alliances due to their efficiency in organizing for collective action in territorial competition.

---

environments (Steward 1955; Winterhalder and Smith 2000; Odling-Smee et al. 2003; Kappeler et al. 2003). Technologies of production and competition therefore play important roles in driving transitions and shaping dynamic equilibria.

A shift to reliance on high-quality agricultural land in the Holocene is likely to have increased the effective variance (patchiness) in the productivity of sites, increasing territoriality, alliance formation, and hierarchical territorial alliances. In other words, intensive agriculture transformed the economic value of territorial resources and created the crucible that produced the chiefdoms and states characteristic of the middle and later Holocene. Sedentism and higher population densities may have also reduced the costs of alliance formation, enforcement, and cooperation in ways that reinforced these outcomes. Mobile foragers relying on widely or unpredictably distributed plant and animal resources, on the other hand, have been less likely to go down this road to increasing sociopolitical complexity.

The dynamics and outcomes this model bear analogy with the depictions of Leach's *Political Systems of Highland Burma* (1973) and Scott's *The Art of Not Being Governed* (2010). In light of the empirical patterns Leach and Scott describe in Southeast Asia, the four environments in Figure 3 could be stylistically interpreted as: (A) and (B) inter-riverine forest tracts utilized by relatively egalitarian and acephalous foragers and farmers; (C) highland valleys suitable for shifting cultivation and agriculture inhabited by a mix of anarchic and hierarchical communities; and (D) productive alluvial valleys dominated by intensive agriculture and large-scale hierarchical polities.

In the Mississippian southeast, fortified villages cultivated crops on well-drained, sandy loam soils that were regularly renewed by inundation. Hudson (1976) proposed that once the best riverine soils came under cultivation, communities became environmentally circumscribed and joined together "beneath the mantle of a chief powerful enough to effect harmony... [and] to stand more strongly against mutual outside foes." The scale of integration and extent of sociopolitical complexity appear to have waxed and waned with the productivity and spatial extent of contiguous arable land (Munoz et al. 2015).

The development of the Northwest Coast hierarchical complex appears to have required millennia (Ames 1994; Matson and Coupland 1994), suggesting relatively slow rates of accumulation of the cultural traits required for the development of coherent hierarchies. Rapid social transitions following the introduction of new resources, technologies, or cultural models are also common in distant and recent history. Small-scale communities confronted by colonization and acculturation in the last 500 years have often moved toward more hierarchical and unequal social structures (Steward 1938; Hämäläinen 2009) and subordination to imperial or national hierarchies (Lee and Daly 2004; Hooper et al. 2014).

Contrary examples of reductions in hierarchy are also well documented (Currie et al. 2010; Scott 2010). Contemporary groups such as the Sungusungu of East Africa and the Kuna of Panama appear to gain from the complementarities between hierarchical organization and practices that limit the ability of leaders to advance personal gain over community interests (Paciotti and Borgerhoff Mulder 2004; Howe 2002; Kim Hill, personal communication). The theory predicts that hierarchical institutions provide the greatest benefits for their constituents when they coevolve with behaviors limiting their power and cost (Bowles 2012). Hierarchical institutions are constrained and can be replaced when less hierarchical forms of organization efficiently provide the same functions. &

REFERENCES CITED

Ames, Kenneth M.
1994   The Northwest Coast: Complex Hunter-Gatherers, Ecology, and Social Evolution. *Annual Review of Anthropology* 23:209–229.

Bates, Douglas, Martin Maechler, Ben Bolker, and Steven Walker
2013   lme4: Linear mixed-effects models using Eigen and S4. The spatial landscape of the model is R package version 1.1-8.

Boehm, Christopher
2001   *Hierarchy in the Forest: The Evolution of Egalitarian Behavior.* Harvard University Press, Cambridge, Massachusetts.

Boone, James L.   ~127~
1992   Competition, Conflict, and the Development of Social Hierarchies. In *Evolutionary Ecology and Human Behavior*, edited by E. A. Smith and B. Winterhalder. Aldine de Gruyter, Hawthorne, New York.

Bowles, Samuel
2012   Pirates, Potlatches, and Proto-states: The Co-Evolution of Leadership and Solidarity Prior to the Emergence of States. Working Paper, Santa Fe Institute, Santa Fe, New Mexico.

Boyd, Robert, Herbert Gintis, and Samuel Bowles
2010   Coordinated Punishment of Defectors Sustains Cooperation and Can Proliferate When Rare. *Science* 328(5978):617–620.

Brown, Jerram L.
1964   The Evolution of Diversity in Avian Territorial Systems. *Wilson Bulletin* 76:160–169.

Currie, Thomas E., Simon J. Greenhill, Russell D. Gray, Toshikazu Hasegawa, and Ruth Mace
2010   Rise and Fall of Political Complexity in Island South-East Asia and the Pacific. *Nature* 467(7317):801–804.

Dyson-Hudson, Rada, and Eric Alden Smith
1978   Human Territoriality: An Ecological Reassessment. *American Anthropologist* 73:77–95.

Gintis, Herbert
2009   *Game Theory Evolving.* 2nd ed. Princeton University Press, Princeton, New Jersey.

Gunther, Erna
1972   *Indian Life on the Northwest Coast of North America as Seen by the Early Explorers and Fur Traders During the Last Decades of the Eighteenth Century*. University of Chicago Press, Chicago.

Hämäläinen, Pekka
2009   *The Comanche Empire*. Yale University Press, New Haven, Connecticut.

Handcock, Mark S.
2015   reldist: Relative distribution methods. R Package Version 1.6-4.

Hirshleifer, Jack.
2001   *The Dark Side of the Force: Economic Foundations of Conflict Theory*. Cambridge University Press, Cambridge.

Hooper, Paul L., Michael Gurven, and Hillard S. Kaplan
2014   Social and Economic Underpinnings of Human Biodemography. In *Sociality, Hierarchy, Health: Comparative Biodemography*, edited by M. Weinstein and M. A. Lane, pp. 169–196. National Academies, Washington, DC.

Hooper, Paul L., Hillard S. Kaplan, and James L. Boone
2010   A Theory of Leadership in Human Cooperative Groups. *Journal of Theoretical Biology* 265(4):633–646.

Howe, James
2002   *The Kuna Gathering: Contemporary Village Politics in Panama*. 2nd ed. Fenestra Books, Tucson, Arizona.

Hudson, Charles
1976   *The Southeastern Indians*. University of Tennessee Press, Knoxville.

Jackson, Matthew O.
2008   *Social and Economic Networks*. Princeton University Press, Princeton, New Jersey.

Kappeler, Peter M., Michael E. Pereira, and Carel P. van Schaik
2003   Primate Life Histories and Socioecology. In *Primate Life Histories and Socioecology*, edited by P. M. Kappeler and M. E. Pereira, pp. 1–20. University of Chicago Press, Chicago.

Kennett, Douglas J.
2005   *The Island Chumash*. University of California Press, Berkeley.

Leach, Edmund R.
1973   *Political Systems of Highland Burma*. Athlone Press, London.

Lee, Richard B., and Richard Daly
2004   *The Cambridge Encyclopedia of Hunters and Gatherers.*
Cambridge University Press, Cambridge.

Matson, Richard Ghia, and Gary Coupland
1994   *The Prehistory of the Northwest Coast.* Left Coast Press,
Walnut Creek, California.

Maynard Smith, John, and George R. Price
1973   The Logic of Animal Conflict. *Nature* 246.

McElreath, Richard, and Robert Boyd
2008   *Mathematical Models of Social Evolution.* University of
Chicago Press, Chicago.

Munoz, Samuel E., Kristine E. Gruley, Ashtin Massie, David A. Fike, Sissel
Schroeder, and John W. Williams
2015   "Cahokia's Emergence and Decline Coincided with Shifts of
Flood Frequency on the Mississippi River. *Proceedings of the National
Academy of Sciences* 112(20):6319–6324.

Odling-Smee, John F., Kevin N. Laland, and Marcus W. Feldman.
2003   *Niche Construction: The Neglected Process in Evolution.*
Princeton University Press, Princeton, New Jersey.

Paciotti, Brian, and Monique Borgerhoff Mulder
2004   Sungusungu: The Role of Preexisting and Evolving Social
Institutions among Tanzanian Vigilante Organizations. *Human
Organization* 63(1):112–124.

R Development Core Team
2008   R: A language and environment for statistical computing. R
Foundation for Statistical Computing, Vienna.

Rick, Torben C., Jon M. Erlandson, René L. Vellanoweth, and Todd J.
Braje
2005   From Pleistocene Mariners to Complex Hunter-Gatherers:
The Archaeology of the California Channel Islands. *Journal of World
Prehistory* 19(3):169–228.

Scott, James C.
2010   *The Art of Not Being Governed: An Anarchist History
of Upland Southeast Asia.* Yale University Press, New Haven,
Connecticut.

Smith, Eric Alden, and Jung-Kyoo Choi
2007   The Emergence of Inequality in Small-Scale Societies:
Simple Scenarios and Agent-Based Simulations. In *The Model-Based
Archaeology of Socio-Natural Systems,* edited by S. van der Leeuw and
T. Kohler. SAR Press, Santa Fe, New Mexico.

Smith, Eric Alden, Kim Hill, Frank Marlowe, David Nolin, Polly
    Wiessner, Michael Gurven, Samuel Bowles, Monique Borgerhoff
    Mulder, Tom Hertz, and Adrian Bell
    2010   Wealth Transmission and Inequality Among Hunter-
    Gatherers. *Current Anthropology* 51:19–33.

Steward, Julian H.
    1938   *Basin-Plateau Aboriginal Sociopolitical Groups.* Bureau of
    American Ethnology Bulletin 120, Washington, DC.

    1955   The Concept and Method of Cultural Ecology. In *The
    Theory of Culture Change: The Methodology of Multilinear Evolution,*
    edited by J. H. Steward, pp. 30–42. University of Illinois Press,
    Urbana.

Trigger, Bruce G.
    2003   *Understanding Early Civilizations: A Comparative Study.*
    Cambridge University Press, Cambridge.

Wilensky, Uri
    1999   NetLogo. Center for Connected Learning and Computer-
    Based Modeling, Northwestern University, Evanston, Illinois.

Winterhalder, Bruce, and Eric Alden Smith
    2000   Analyzing Adaptive Strategies: Human Behavioral Ecology
    at Twenty-Five. *Evolutionary Anthropology* 9(2):51–72.

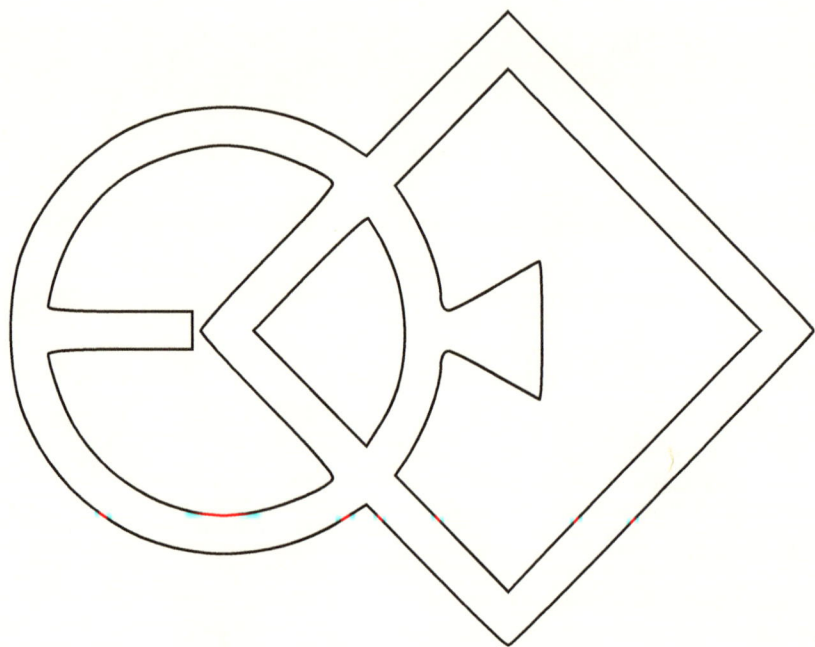

ॐ

SOCIOPOLITICAL EVOLUTION IN MIDRANGE SOCIETIES: THE PREHISPANIC PUEBLO CASE

Timothy A. Kohler, Washington State University and Santa Fe Institute
Stefani A. Crabtree, Pennsylvania State University
R. Kyle Bocinsky, Washington State University
and Paul L. Hooper, Santa Fe Institute

Here we revisit, with new data, tools, and theory, *the* classic problems engaging social and political theorists since at least the time of Hobbes (*Leviathan*, 1651): how and why, over the last few thousand years, did the relatively egalitarian foraging bands of our deep prehistory give way to larger-scale societies marked by obvious inequalities in power and wealth? Although the end points of this process may be fairly clear, what's in the middle remains a muddle. We develop our approach with reference to a specific historical trajectory, yet we suspect this model represents a common path to sociopolitical complexity in the absence of direct competition with larger, more hierarchical groups. Our proof-of-concept model reproduces important aspects of patterns in settlement and conflict seen in the central Mesa Verde region of the Pueblo Southwest in the last half of the first millennium and the early second millennium AD.

Outputs from this model, however, do not map very well into the taxonomies developed by neo-evolutionary studies of the mid-twentieth century (e.g., Fried 1967; Service 1962). This is a little troubling, but on the other hand archaeologists have often lamented the poor fit of concepts like "chiefdom" or "stratified society" to what they see as the facts on the ground in the later pre-Hispanic Southwest (e.g., Haas et al. 1994). In any case, we are more interested here in process than taxonomy.

Clarity, though, requires some vocabulary. The model

developed here recognizes three basic kinds of groups beyond the household: simple nonhierarchical groups, simple hierarchical groups, and complex hierarchical groups composed of multiple groups. We build an evolving ecosystem of households within these three types of groups that has no preordained end point. What hap-

How is the natural reluctance of people to give up their political autonomy (or to contribute to the public good) overcome (or minimized) in increasingly hierarchical groups?

pens in any specific run is strongly conditioned by structural factors such as resource distribution and abundance, population sizes of groups, and distribution of groups; "history" (here, any factor that structures subsequent development) also plays an important role. In this model the households within a group can be expected to have only modest internal differences in power or wealth. This is in keeping with the characteristics of the local archaeological record, which seems to attest to societies with a bias toward relative equality at least through the Pueblo I period (ending ~AD 890; Kohler and Higgins 2016). After that time, during the period of maximum local influence from Chaco and Chaco-derived societies to the south (~AD 1060–1140), there is good evidence for increasing differentiation of wealth between households (using house size as a proxy for wealth; Kohler and Ellyson 2018). Little is known about differences in wealth between groups, although the simulation reported here predicts that such differences might be important after roughly AD 1000. We show that complex groups might become large enough to dominate an area equal in size to the area we simulate. Pauses in conflict seen in the archaeological record of this area therefore might be explainable by suppression of conflict within such a group.

The approach we take embodies both of the pathways by which sociopolitical complexity may increase, identified by Hobbes 350 years ago. He believed that a basic human motivation to acquire power could easily result in continual struggles for supremacy and possessions among individuals that in turn lead to a "solitary, poor, nasty, brutish, and short" existence in the chaos of individuals freely exercising their natural rights in close proximity. But of course we have not been willing victims of these circumstances. Hobbes said that individuals can escape this outcome by abdicating a portion of their individual rights to a sovereign power (Leviathan) that "may use the strength and means of them all...for their peace and common defense" (Hobbes [1651]1957:112):

> The attaining to this sovereign power, is by two ways. One, by natural force; as when a man maketh his children, to submit themselves, and their children to his government, as being able to destroy them if they refuse; or by war subdueth his enemies to his will, giving them their lives on that condition. The other, is when men agree amongst themselves, to submit to some man, or assembly of men, voluntarily, on confidence to be protected by him against all others. This latter, may be called a political commonwealth, or commonwealth by *institution*; and the former, a commonwealth by *acquisition*. (112–113)

Hobbes's two alternatives for the emergence of leaders continue to structure debate in both political theory and anthropology on how sociopolitical complexity may increase. Conflict theorists (e.g., Carneiro 1970) emphasize pathways in which hierarchy is imposed. More functionally minded scholars (e.g., Johnson 1978) envision managerial elites voluntarily supported for the good works they achieve.

These two competing positions have been able to survive only because there is some support for each. This model regards the emergence of political hierarchy as a process in which voluntaristic, small-scale "commonwealths by institution" (simple hierarchical groups) may become nested within larger-scale "commonwealths by acquisition" with the formation of complex groups. This happens as simple groups grow in population and come into competition with other groups of similar scale. Simple groups with no leaders, however, are limited in size by their inability to coordinate their activities. Thus through time, the largest groups in the model may first be simple nonhierarchical groups, but as group and regional populations grow, simple groups with leaders gain an advantage and displace many of the simple nonhierarchical groups. Eventually, simple hierarchical groups come into conflict with each other, and, typically, larger groups subsume smaller groups by force or negotiation, forming complex groups composed of two or more simple groups.

~136~

The model we propose assumes that individuals have long ago found ways to cooperate within families (households). Simple groups, internally united in our model by ties of kinship and possibly success in provisioning public goods, are allowed to grow until they encounter a numeric threshold (our GROUP_SIZE parameter) that corresponds notionally to the approximate scale of a clan or a small group of related clans (phratry). That these groups are fairly small is no accident. Mancur Olson, one of the original formulators of public goods theory, noted that small groups are expected to deliver optimal amounts of a collective good better than large groups (Olson 1971:35). If a group is so large that each individual's actions do not make a noticeable contribution to the group, Olson argued, an individual will have no incentive to contribute unless there are "selective" positive or negative incentives (Olson 1971:50–51). Thus, the simple leaderless groups we model are fairly small, with further growth only made possible by the action of leaders who (1) provide selective negative incentives against those who fail to cooperate, and thereby (2) allow for a positive return to group size through repeated collective action.[1]

Rates for the cooperative and competitive processes in the model (and, we believe, in the world) are spatially and temporally variable, depending on the underlying productivity of the landscapes in which they are embedded, as well as the spatial, demographic, and organizational characteristics of groups. Moreover, these processes have an inevitable historical dimension (path dependency), given their evolutionary character (in which future actions are partially conditioned by present circumstances) and some randomness in various processes.

~137~

Any attempt to endogenize rates of population growth and productivity, as we do here, must begin with realistic modeling of resource landscapes. We implement a model of self-regarding households interacting over these resources within the model for group formation and evolution described below. This mode of inquiry minimizes traditional concerns such as "do social relations prevail over technological and environmental considerations, or do these latter 'ecological' domains pose primary constraints on the evolution of political systems and social structures?" (Upham 1990:9). Instead, we are able to ask, how do social and ecological dynamics interact in the evolution of political systems?

Another classic concern that we implicitly address with this approach is the notion of resistance: how is the natural reluctance of people to give up their political autonomy (or to contribute to the public good) overcome (or minimized) in increasingly hierarchical

---

[1] Since this chapter was completed, we have expanded portions of this analysis to consider the relationship between the polities simulated here and those visible on the ground in the northern Southwest as the local manifestations of the "Chaco phenomenon" (Crabtree et al. 2017). The code for that version, which is similar to that reported here except that it adds revolt and fission, is available at: https://github.com/crowcanyon/vep_sim_beyondhooperville.

groups? The groups we model are made up of actors with differing inherited proclivities for degrees of prosocial vs. self-regarding action. The variable success of these differing strategies through

---

Complexity *is an unfortunate term because its inversion is* simplicity, *but no known society of* Homo sapiens *is (or has ever been) simple.*

---

time is determined by running the model, not by decisions we make in advance of the modeling, though our choices of plausible parameter values (especially for the public goods game) do influence the success of the various strategies. We propose that many such classic dilemmas of sociopolitical theorizing will dissolve as specific historical instances are modeled with adequate endogeneity. Is it resource stress or resource abundance that is most likely to lead to institutionalized inequality? Are polities inherently born of conflict or cooperation? Which came first, control over resources or social power? Many such questions turn out to be coevolutionary in nature, and small initial differences may become substantially magnified through nonlinear interactions.

Testing any model's explanatory power requires a concrete empirical context. Building on the models developed in Hooper et al. (2010) and Kohler et al. (2012), the model in this chapter is implemented on an 1,800 km² landscape resembling that of Southwest Colorado from AD 600–1280, described by Ortman et al. (2012). This is also called the Village Ecodynamics Project (VEP) I area.

What Do We Mean by Sociopolitical Complexity?

Complexity is an unfortunate term because its inversion is *simplicity*, but no known society of *Homo sapiens* is (or has ever been) simple: "The notion of complexity in anthropology makes sense only in making typological distinctions of scale and hierarchies of decision making, not with regard to the number of interactions or relationships among constituent agents or groups in a society" (Clark 2002). Nor are even small-scale human groups completely egalitarian, since they typically support socially defined distinctions along the lines of age, gender, size, ability, and kinship (Feinman 1995:256–257; von Rueden et al. 2014; Wiessner 2002:251). The landscape on which "the road to inequality" is built is thus strewn with abundant raw materials.

~139~

As Drennan and Peterson (2012) point out, the processes of sociopolitical evolution have been variable enough that it is difficult to agree on a general definition of sociopolitical complexity that adequately describes the available cases. With reference to the midrange societies in the US Southwest that concern us here, Lightfoot and Upham (1989) defined sociopolitical complexity as including the development of hierarchical decision-making organizations, the presence of status differentiation, and the rise of inequality that limits access to economic resources and ritual information. Of course, objections can be raised to any one of these criteria. Netting (1990), for example, has demonstrated that the last character may be present among intensive cultivators in acephalous communities, and Braun (1990) notes that some delegation of authority occurs in nonhierarchical communities. Clearly it is easier to define processes of *increasing* sociopolitical complexity and demographic scale in specific historical trajectories, as we do here, than to define invariants across cultural traditions and regions.

Recent Approaches to Understanding Emergence of Leadership

In the last third of the twentieth century, North American archaeologists were primarily concerned with correctly identifying complexity when they saw it and in weighing the role of factors such as craft specialization, sedentism, storage, long-distance exchange, population increase, and so forth in causing sociopolitical change toward greater economic or social inequality and more hierarchy in decision-making (e.g., Plog 1990). Although many of these researchers criticized aspects of midcentury neo-evolutionary syntheses, on the whole there was considerable continuity with the way the problem of sociopolitical evolution was conceptualized and addressed. Rosenberg (2009:24) has characterized the dominant approach as "progressive transformationalism."

Some recent treatments of these issues, however, propose a radical break with this tradition via construction of formal models that focus on

- how within-group cooperation can be achieved and maintained (known in the political science literature as the collective-action problem) given a rational-actor model. The importance of punishment and reward in particular is becoming more obvious (Boyd and Richerson 1992; Hooper et al. 2010), and not just within human groups. Flack et al. (2013) show that punishment is key to within-group cohesion in groups of pigtailed macaques, and that suppressing policing mechanisms destabilizes social networks;
- the structure and size of groups and the metapopulation in which groups reside and interact (Hamilton et al. 2007);
- intergroup competition and conflict (rarely mentioned by

~I40~

southwestern archaeologists until the mid-1990s, though
see Lightfoot and Upham [1989]);

- evolutionary dynamics, often involving strategic games to
  drive them (see Stanish [2009] for a discussion of game
  theory in relation to sociopolitical evolution);
- an appreciation that hierarchy may confer advantages
  within groups for coordination, efficiencies in information transmission, or reduction of environmental uncertainty (Flack et al. 2013);
- the recognition that hierarchy may be able to spread via
  demographic effects resulting from uncoupling resource
  availability from reproduction (Rogers et al. 2011);
- suggestions from numerous quarters that human social
  systems may become more complex in a variety of ways
  that do not necessarily involve greatly increased centralization and hierarchy (e.g., Hooper et al. 2015; Mezza-Garcia
  et al. 2014);
- how prosocial tendencies (such as a willingness to die
  for one's group) could evolve, on long time scales, in
  populations of self-regarding individuals (Bowles and
  Gintis 2011); and
- the general rise of a complex adaptive systems perspective (Holland 2014; Kohler 2012a), with its attention
  to heterogeneity across scales and emergent properties,
  such as leadership and institutions, whose description
  and analysis typically combine agent-based modeling
  with analytical approaches. These approaches can bring
  in "big picture" considerations frequently missing in
  post-neo-evolutionary applications of evolutionary theory
  by archaeologists (Bettinger 2009) while honoring the
  microevolutionary processes on which such archaeologists
  have focused.

These new directions have inspired us to consider three specific
design requirements for the present model. First, many small-scale
human groups may be sufficiently stable and strongly enough dif-
ferentiated from other groups, genetically or culturally, to support

group selection (Henrich 2004). Contrasting selection pressures may thus act on the level of the individual and the group; for example, "selfishness beats altruism within groups. Altruistic groups beat selfish groups" (Wilson and Wilson 2007:335). Between-group competition is a main motor for increased social complexity and inequality (Flannery and Marcus 2012:473).

Second, the initial steps toward hierarchy and power inequalities must be very small and acceptable within a tradition of egalitarianism typical of small-scale societies. This is likely to involve voluntary participation that benefits everyone in the group in some way. Rosenberg (2009:37–40; see also Feinman [1995:263]) has suggested that internal peacekeeping (conflict resolution) provides a legitimate, "primitive," general social role meeting this requirement. Explanations considering the local contexts in which leadership first becomes evident in the archaeological record referenced in our model (Pueblo I villages, mid- to late AD 700s) have suggested that lineage heads could have met an "original social purpose for leadership" by organizing the increasingly long-distance hunts required to return deer to the villages in an increasingly game-depressed landscape (Kohler and Reed 2011) and distributing the returns in a fair manner (i.e., conflict prevention).

Third, defense or predation against other groups is a public good, conferring advantages on groups at a cost to the participant (Bowles 2009; Turchin and Gavrilets 2009). "Warfare is a [particularly] high-stakes form of cooperation" (Mathew and Boyd 2013:58). Although the altruist as warrior is paradigmatic, a "willingness to take mortal risks as a fighter is not the only form of altruism that contributes to prevailing in intergroup contests; more altruistic and hence more cooperative groups may be more productive and sustain healthier, stronger, or more numerous members, for example, or make more effective use of information" (Bowles 2009:1294).

To reflect these insights, we can develop a model that (1) allows a multilevel selection dynamic in which social strategies within groups can evolve; (2) builds complexity from a starting point of voluntary participation, naturally modeled as a public goods game; (3) allows for policing/punishment to maintain within-group cooperation until such point as (4) between-group competition, including conflict, allows leadership (or groups with leaders) to take on more coercive properties. We implement this approach in a specific spatially and temporally heterogeneous environment, allowing us to explicitly evaluate the realism of the ~143~ dynamics generated by the simulation.

A Verbal Description of the Model

THE BASE AUTONOMOUS-HOUSEHOLD-ECOLOGY MODEL[2]

The simulation begins in "AD 600" by randomly seeding 200 households on a virtual landscape that we have endowed, to the best of our ability, with realistic levels of four resources (water, woody fuels, three species of huntable prey, and potential maize fields) whose spatial distribution varies according to edaphic factors and whose temporal distribution varies in accordance with tree-ring-proxied climates in our study area, which affect resource growth rates (Johnson and Kohler 2012; Kohler 2012b; Kolm and Smith 2012). Household activities for this base model (referred to as "Village" and described by Kohler 2012c) are incorporated in the current simulation. In brief, households myopically and approximately minimize their caloric costs for obtaining adequate supplies of all these resources through central-place foraging, prey switching, labor intensification, and household relocation, as befits their local circumstances and possibilities.

---

2 A Swarm implementation of an earlier version of this model is deposited in OpenABM (www.openabm.org/model/2518/version/2/view).

TABLE 1 Agent types and approximate payoffs related to their participation in the public goods game.[1]

Type	Approximate payoffs
NH.ALLC (nonhierarchic, always cooperate)	$V (ALLC \mid p, q, r) = (1 + (p + q + Qr)(n - 1))b/(n - c)$ where Q represents probability that at least one other member of the group is a monitor $(Q = 1 - (1 - q)^{n-1})$.
NH.MM (nonhierarchic, mutual monitor)	$V (MM \mid p, q, r) = [(1 + (p + q + r)(n - 1))b/(n - c - c_m(n - 1)] - rc_s(n - 1)$
NH.RC (nonhierarchic, reluctant cooperator)	$V (RC \mid p, q, r) = [(Q + (p + q + Qr)(n - 1))b/(n - Qc)] - sq(n - 1)$
H.ALLC.T.L (hierarchic, always cooperate, taxpayer, leader)	$V(L \mid u,v) = uvtbn - c_m n - (1 - u)c_s n - (1 - v)\hat{c}sn + (tb - c_m)n$
H.ALLC.T.UL (hierarchic, always cooperate, taxpayer, not leader)	$V (H.ALLC.T \mid u, v) = [1 + u(n - 1)](1 - t)b/n - c + [(1 - t)b - c]$
H.ALLC.RT.L (hierarchic, always cooperate, reluctant taxpayer, leader)	Same as for H.ALLC.T.L
H.ALLC.RT.UL (hierarchic, always cooperate, reluctant taxpayer, not leader)	$V (H.ALLC.RT \mid u, v) = [1 + u(n - 1)])b/n - c - \hat{s} + [(1 - t)b - c]$
H.RC.T.L (hierarchic, reluctant cooperator, taxpayer, leader)	Same as for H.ALLC.T.L.
H.RC.T.UL (hierarchic, reluctant cooperator, taxpayer, not leader)	$V (H.RC.T \mid u, v) = u(n - 1)(1 - t)b/n - s + [(1 - t)b - c]$
H.RC.RT.L (hierarchic, reluctant cooperator, reluctant taxpayer, leader)	Same as for H.ALLC.T.L
H.RC.RT.UL hierarchic, reluctant cooperator, reluctant taxpayer, not leader)	$V (H.RC.RT \mid u, v) = u(n - 1)b/n - s - \hat{s} + [(1 - t)b - c]$

[1] Payoffs are approximate since it cannot be known in general whether reluctant cooperators (in both group types) or reluctant taxpayers (in the hierarchically inclined groups) will need to be punished in any given year. Payoffs to leaders refer to agents actually acting as leaders; potential (latent) leaders receive payoffs appropriate to their actions as regular members of their hierarchical group. When agents are

We track household composition (number of members, sexes, and ages), and household requirements scale according to household size and composition. Households move on formation, and also when their current location becomes untenable because of declining resource yields or growing household size. Since a number of households initially land in poor areas, a decrease in household number in the first three to four years of the simulation is typical. In the simulations reported here, we allow households to engage in time-delayed reciprocal exchanges of maize for maize and meat for meat, both with close kin ("generalized reciprocity") and near neighbors of good standing ("balanced reciprocity") (Crabtree 2015; Kobti 2012). Suppressing exchange would slightly decrease global household numbers and degree of aggregation in the simulations (Crabtree 2015).

~145~

EVOLUTIONARY PUBLIC GOODS GAME

While retaining all the behaviors represented in the base simulation, we add a number of features enabling us to grow groups and leaders. Inspired by Hooper et al. (2010), we instantiate three social strategies typifying individuals who prefer to live in nonhierarchical groups and eight social strategies typifying agents willing to live in hierarchical groups (Table 1). Initially these strategies are randomly distributed among the members of each household, but as new households are formed (via marriage of a daughter) the new household assumes the social strategy of the wife's mother if she is alive; otherwise they take on the preferences of the wife's father.

Once a year, all households play a public goods game within their group. In the general game, households put a certain amount

---

nonhierarchically inclined, p = fraction of pure cooperators; q = fraction of monitors; r = fraction of reluctant cooperators; $p + q + r = 1$. When agents are hierarchically inclined, u = fraction of pure cooperators; v = fraction of willing taxpayers; y = fraction of individuals willing to lead. From Kohler et al. (2012), modified from Hooper et al. (2010).

of a resource (maize in our case) into a public fund. The amount in this fund is multiplied by a factor representing the return on the public good, and then is redistributed equally to all group members (we call this augmented amount the *benefit* of the public good). If all households contribute to the public good, each gets a good return on its investment. If just a few households in a group do not contribute, those defectors not only keep what they should have donated but share in the return accruing to each household in the group. Thus each household has a temptation to defect. Unfortunately, in fact, the unique Nash equilibrium is for all households to defect (Capraro 2013:5). Nonhierarchical groups may contain one or more "mutual monitors" who monitor and punish defectors at some cost to themselves (Table 1). Hierarchical groups will contain a leader who fulfills these same functions and who is reimbursed through a tax. Such leaders can be very roughly conceptualized as "big men" with no coercive power except within the limited domain where the group voluntarily grants it. Members of hierarchical groups must pay this tax *and* contribute to the public good. Obviously, the hierarchical preference will thrive only when the tax plus the contribution to the public good is less than the return on that good. In the model, and we believe in the world, getting viable rewards from the public goods game requires close vigilance and occasional punishment.

Households can exist in three states: thriving (2), just getting by (1), and perishing (0). The default value is 2, but this gets lowered to 1 if the amount of maize in storage is less than that needed for the current year plus that expected to be needed for the following year *or* if the maize just harvested is less than next year's anticipated needs. Households in state 1 reproduce according to a life table that provides for an approximately stable global population.

We define a parameter "STATE_GOOD" that determines the degree to which natality and mortality are affected by the household's state. When STATE_GOOD = 1 (the value we apply here), the probabilities of giving birth are incremented by 10 percent for women in a household in state 2 (from probabilities in an empirically derived life table; Kohler 2012c:68; Weiss 1973:156), and the probabilities of dying are decremented by 10 percent for members of that household. A household's hierarchical preference and strategy for playing the public goods game affect its maize storage and perhaps its state, and may therefore increase, or decrease, its relative number of offspring, who inherit the parent's strategy, providing a slow evolutionary dynamic to strategy change in the population. Optionally, but implemented here, we define a faster social learning dynamic in which agents emulate the propensity of the "richest" household (that with the most storage) to work in a hierarchical setting, though not its other behaviors related to the public goods game. This is a model of indirect bias as defined by Boyd and Richerson (1985:241–259).

~147~

Up to this point, the model corresponds to that implemented and analyzed by Kohler et al. (2012). Among other findings, Kohler et al. (2012) reported that most households initially prefer to live in nonhierarchical groups, but as those groups grow in size (which happens first in the most productive regions), "mutual monitors"—who begin to pay more for these activities than they receive as their share in the public good—are at a competitive disadvantage compared to other agent types, and decline in frequency. As this happens, nonhierarchical group members will receive less return from the public good as more and more members of their group fail to contribute and are not punished for this failure.

Conversely, members of hierarchical groups will not do very well when their groups are small but will prosper more as they increase in size. Taxes paid to support a leader (who punishes those failing to contribute to the public good) ensure that everyone contributes to the public good. Accordingly, hierarchical groups continue to grow in size and dominate the most productive areas. Nonhierarchical groups remain small and tend to dominate just those areas with poor production.

FOUR IMPROVEMENTS OVER THE PREVIOUS MODEL

Kohler et al. (2012:12–24) report more details on the implementation of the public goods game than we have space to review. Below, we describe four modifications to that model that address its main weaknesses as we see them.

First, whereas in the earlier work groups were formed by assignment of nearby households, in the work reported here households track their lineage and grow groups based on kinship. These lineages are the original "groups" in the simulation, and grow (or not) according to how well their constituent households thrive on a variable landscape (which is in part determined by the social strategies of the households). The founding households seeded on the landscape are assigned unique lineage identifiers that are inherited matrilineally by daughter households.[3] No new lineage identifiers are created during the simulation, nor do we model any immigration, so each surviving agent household tracks its heritage back to its founding household. As groups grow, they may fission if they reach maximal group size. This can be considered a "span of control" measure; it forms an assumption as to how big a group can become and still act as a single (simple) group. For the Hopi, Levy (1992:20) describes cases where groups exceed in size the carrying capacity of farmland, in which case extended families bud off to form new groups. Alternatively, fission could be due to scalar

~148~

---

[3] We take no position here on whether the kinship system in the world we model is unilineal and, if so, whether it employed a matrilineal/matrilocal or patrilineal/patrilocal bias. In the model as it presently exists, this distinction between biases is irrelevant, except that we expect a faster pruning of patrilines than matrilines from the population because only males die in warfare. In the real world, however, these systems do have important differences, for example with respect to rates of internal versus external warfare (see Ember and Ember 1971).

stress (e.g., Johnson 1982). The model makes no assumptions about which is the correct mechanism.

Second, a group now decides whether to be hierarchical or nonhierarchical based on the majority preference of its constituent households. In earlier simulations, groups were formed only of households with the same preferences (Kohler et al. 2012:13). This

---

When a group has a frustration that hurts, it has the opportunity to relieve frustration by tendering an offer to subsume another group as its subordinate (to "merge" and thus form a complex group) or, if that offer is rejected, to fight.

~149~

---

led to the strong selective dynamic noted by Hooper et al. (2010) and Kohler et al. (2012) whereby larger groups preferenced hierarchical agents. Our groups are now determined by kinship, so while kin will tend to have similar preferences through inheritance, there are often groups with mixed preferences. All households have *all* behavioral preferences required to play either the hierarchical or nonhierarchical public goods game; for example, a household with a hierarchical preference may be in a nonhierarchical group, in which case its hierarchical-type preferences (willingness to be a leader, tax rate, and whether or not they are a reluctant taxpayer) will not be activated, while its nonhierarchical preferences (willingness to be a mutual monitor) will be expressed. As we discuss below, this dynamic of majority-rules play and socially dependent expression of preferences has a large impact on the resilience of specific—and even nonadaptive—preferences in agent populations.

Third, groups are now territorial, in contrast to groups in the earlier simulation that could intermingle with no restrictions. Not only is there a great deal of evidence suggestive of territoriality from spatial distributions of dwellings in the study area (e.g., Reese 2014; Varien 1999) but defended claims to territory also figure prominently in most explanatory models for sociopolitical evolution (e.g., Boone 1992; Gibson 2008; Hooper et al., chapter 5 in this volume; Maine 1861; Smith and Choi 2007).

Fourth, we add two mechanisms—merging and fighting—by which two or more simple groups may form a complex group. The importance of intergroup competition in current theory has already been noted; Kohler et al. (2014) summarize and analyze evidence for violence through time in the study area referenced here. We now provide more detail on each of these modifications.

TERRITORIALITY, MERGING, AND WARFARE

Groups in the model are corporate: they maintain and defend claims to the core portion of their territory used for growing maize. (They do not own or defend the larger territories usually necessary to acquire other resources.) As some of the initial 200 groups prosper and grow on the landscape, a convex-hull polygon is drawn around their member households, and no other group is allowed to plant within or move into that polygon. As daughter households bud off the original household, the polygon grows to encompass those daughter households and their fields. Currently, fields must be either in the same 200-m cell where the household resides or in one of its eight neighbors (its Moore neighborhood).

At the beginning of the simulation, 200 households are seeded randomly on the landscape and told to move to the best available location within the MOVE_RADIUS parameter (40 cells, or 8 km), subject to the rules governing territoriality noted above. Households then reevaluate their locations annually and attempt to move if their anticipated needs are not likely to be met. Not all desired moves are allowed, however. Cells are disallowed that are in other groups' territories, that would result in overlap of group territories, or that would require crossing another group's territory to access. Each time a household cannot move to a cell to which it would like to move, it tracks the group that impeded its move. We call these "frustrations." If a household cannot move to *any* cell that is higher ranking than its home cell, its group records this as a "frustration that hurts." Frustrations that hurt can lead to merging or warfare. Frustrations (including those that hurt) are tracked at the group level.

When a group has a frustration that hurts, it has the opportunity to relieve frustration by tendering an offer to subsume another ~151~ group as its subordinate (to "merge" and thus form a complex group) or, if that offer is rejected, to fight. Each group archives a list of groups that have frustrated it. This list is sorted according to a function that considers the distance between the two groups and the quantity of frustrations incurred. The focal group will then iterate through its frustrations, calculating its likelihood of winning a battle against each group. Specifically, the focal group will compare its likelihood of winning battle against a random number between 0 and 1. If the random number is less than its probability of winning in battle, the focal group will decide to tender an offer of merging and potentially fighting.

Let's call the aggressing group *m* and the defending group *n*. Group *m* will always first tender an offer of merging. Group *n* will then calculate its probability of winning a potential fight (see the section "Warfare: Stochastic Lanchester Laws," below); this proportion is compared against a random number between 0 and 1 as above, and group *n* accepts the merger if the random number is less than or equal to its own probability of winning (p_n). In that case, group *n* will become subordinate to group *m*, forming a complex group. Smaller groups are more likely to accept an offer of merger than larger groups, whereas evenly matched groups have even odds of accepting or rejecting an offer to merge. Each group can accurately estimate the size of opposing groups. Simple groups within complex groups will be able to count some warriors from their larger groups in these size estimates (see "Warfare: Stochastic Lanchester Laws"). Numerous ethnographic accounts of groups such as the Shoshone, who would occasionally group together to show their strength to an enemy (D'Azevedo 1986), or the Maori, whose haka dance could allow warriors to show their strength (Ka'ai-Mahuta 2010:106), suggest that this assumption is plausible.

If group n does not accept the offer to merge, then group m decides whether to actually fight. Group m uses the same logic presented above: it calculates the probability of winning a fight against group n (this will be $p_m = 1 - p_n$) and makes a stochastic "decision" based on that probability. Stronger aggressing groups are more likely to decide to fight (Manson and Wrangham 1991).

Should group m decide to fight, the probability of m or n winning is once again calculated (p_m, as before). The outcome of the fight is determined by probabilistically sampling the uniform distribution $[0,1]$ twice (call these d_m and d_n), and comparing each draw to p_m and p_n. If $[p_m \geq d_m$ AND $p_n \geq d_n]$ or $[p_m < d_m$ AND $p_n < d_n]$, the fight is considered a "draw," and each group walks away from the battlefield wounded but not entering into a complex group. Otherwise, the group whose probability of winning met or exceeded its random draw will attempt to subsume the defeated group as a subordinate (see "Complex Groups and Tribute," below).

Regardless of whether a complex group is formed, fights always generate casualties (the removal of a fighter from the battle due to injury or death), a portion of which can result in fatalities. Lanchester showed that in hand-to-hand combat, the number of casualties is approximately equal to the size of the smaller group engaging in battle (Lanchester 1916). We stochastically calculate fatalities for each group independently as a function of the minimum group size $f_{mn} = min(f_m, f_n)$ by simulating f_{mn} coin tosses weighted by a factor s, or the probability that a casualty will result in a fatality. Thus, on average, $2sf_{mn}$ deaths will occur in any given fight between groups of sizes f_m and f_n.

To summarize, merging and fighting occur in the following order. The focal group (1) tenders an offer of merger to the frustrating group; (2) if that offer is rejected, it decides whether to attack the frustrating group; (3) if deciding to attack, it fights the frustrating group (suffering casualties and possibly fatalities); (4) if successful, it subsumes the frustrating group as its subordinate in a complex group, but only if the frustrating group is not already subordinate to another group.[4] A complex group can only have one dominant group at a time but can have multiple subordinate groups. We do not have an upper cap for the number of subordinate groups in a complex group; theoretically, all groups in Village could be contained in one complex group, and in fact this does happen in some of the simulations presented here.

~153~

COMPLEX GROUPS AND TRIBUTE

We call groups in dominant-subordinate relationships "complex groups." They can become much larger than simple groups but are distinctive in two other ways as well: they require their subordinate groups to pay tribute to the dominant group, and they enable some of their constituent groups to call on larger pools of warriors for offense or defense.

Tribute flow is one of the defining characteristics of power in complex societies (Steponaitis 1981); in our model, each subordinate group must pay a tax to its dominant group. Steponaitis proposes that degree of political centralization can be determined

---

[4] It "makes sense" to attack a much smaller group, even if it already has a dominant, because it is likely to wipe some households off the landscape, thus (potentially) relieving frustrations.

from the amount of tribute collected in each hierarchical level and how that tribute flows between the levels in the hierarchy. While he considers the easiest way to measure levels of hierarchy to be the appearance of monumental architecture (which we would consider to be materialized public goods), in this simulation we model flows of tribute in maize, in keeping with Steponaitis's estimates of comestibles and how their flow allows for growth of hierarchy. This is a stylized assumption, which is nonproblematic if labor having an equivalent caloric value was the actual currency employed in our reference context.

Steponaitis assumes that groups consist of producers (farmers) and nonproducers (administrators) and that the job of administrators in a hierarchical society is to ensure the flow of tribute. "In any settlement: (1) the number of producers is directly proportional to the annual yield of that settlement's catchment, minus the food that is allocated as tribute; and (2) the number of non-producers is directly proportional to the amount of tribute in food to which that settlement has access" (Steponaitis 1981:325). As more layers of hierarchy are added, administrative centers keep a portion of tribute from lower levels within the hierarchy, some or all of which is distributed along with the shares of the public good originating within that group itself. Steponaitis calculated that, generally, some 16 percent of produced comestibles was passed up the hierarchy as tribute, although it could be as much as 22 percent in some cases. In our case it seems unlikely that 16–22 percent of individuals would be nonproducers and, in fact, even leaders of hierarchical groups still farm in our simulation. Nevertheless, it seems likely that in the most complex societies in the Pueblo Southwest there was at least some tribute flow—as Mahoney and Kanter (2000:10) argue for the Chacoan system.

In the organizational scenario that Steponaitis envisions, multiple lower-level sites (whose number is limited by the "span of control" variable in Gavrilets et al. [2010]) channel tribute to a higher-level site. If there are sites at a still higher level in the hierarchy, this organization can be scaled accordingly, so that several intermediate-level sites may channel tribute to a paramount site. We note in advance that the model we simulate here is somewhat more likely to form chains of dependency than clusters of sites at the same level, channeling tribute to a single site at the next higher level. Whether this is realistic will be discussed below.

We define β as a tax on a subordinate group's net benefit from the public goods game, and μ as the proportion of the tribute from a subordinate group passed through an intermediate group to a dominant group $(1 - \mu$ therefore being the tax kept on that pass-through). Consider a complex group consisting of four groups $(a \rightarrow b \rightarrow c \rightarrow d)$, where arrows indicate the flow of tribute up the hierarchy from a to b, b to c, and c to d. Let H_i be the net benefit from the public goods game paid to group i, and let μ be a possible compounding factor as tribute moves up the chain. Group a will pay $\beta \cdot H_a$ to group b; group b will pay $\beta \cdot H_b + \mu \cdot \beta \cdot H_a$ to group c; and group c will pay $\beta H_c + \mu (\beta H_b + \mu \cdot \beta H_a)$. This pattern will continue up the chain. More generally, the tribute, T_g, that any group g will pay to their dominant group may be calculated as a function of the benefits from the public goods game of all groups *lower* on the hierarchy than group g and their distance from group g in the hierarchy graph:

$$T_g = \beta \sum_{i=0}^{n} (H_i \cdot \mu^{d_i}) \tag{1}$$

where i indexes the groups in the subordinate neighborhood n of group g, including group g itself, and d_i is the graph distance between group g and group i. Here, following Gavrilets et al. (2010:64), we allow the fixed parameters of β and μ to take on values (0.1|0.5|0.9, Table 2). Gavrilets and colleagues explored

TABLE 2 Parameters varied in this study. Run 39 duplicates run 38, and run 42 duplicates run 41, except for the random number streams they sample. Standard fit is calculated as the negated mean of the standardized Euclidean distances in population and warfare between each run and the empirical record. The highest standard fit (in bold) indicates the best-fit run.

Run	$S*$	Group size**	μ †	β ‡	Type	Standard Fit
1	0.02	50	0.1	0.1	Warfare	0.331
2	0.05	50	0.1	0.1	Warfare	0.576
3	0.02	100	0.1	0.1	Warfare	−0.148
4	0.05	100	0.1	0.1	Warfare	0.084
5	0.02	50	0.1	0.5	Warfare	−2.666
6	0.05	50	0.1	0.5	Warfare	−0.271
7	0.02	100	0.1	0.5	Warfare	0.327
8	0.05	100	0.1	0.5	Warfare	0.117
9	0.02	50	0.1	0.9	Warfare	0.353
10	0.05	50	0.1	0.9	Warfare	0.026
11	0.02	100	0.1	0.9	Warfare	−0.183
12	0.05	100	0.1	0.9	Warfare	0.585
13	0.02	50	0.5	0.1	Warfare	0.056
14	0.05	50	0.5	0.1	Warfare	0.443
15	0.02	100	0.5	0.1	Warfare	−0.179
16	0.05	100	0.5	0.1	Warfare	−0.077
17	0.02	50	0.5	0.5	Warfare	0.335
18	0.05	50	0.5	0.5	Warfare	0.106
19	0.02	100	0.5	0.5	Warfare	0.583
20	0.05	100	0.5	0.5	Warfare	−0.170
21	0.02	50	0.5	0.9	Warfare	0.028
22	0.05	50	0.5	0.9	Warfare	**0.891**
23	0.02	100	0.5	0.9	Warfare	−0.594
24	0.05	100	0.5	0.9	Warfare	0.157
25	0.02	50	0.9	0.1	Warfare	0.266
26	0.05	50	0.9	0.1	Warfare	−0.323

* probability that a casualty will result in a fatality
** how big a group may become before fissioning
† proportion of the tribute from a subordinate group passed through an intermediate group to a dominant group
‡ tax on a subordinate group's net benefit from the public goods game

TABLE 2 *(continued)*

Run	S^*	Group size**	μ †	β ‡	Type	Standard Fit
27	0.02	100	0.9	0.1	Warfare	−0.021
28	0.05	100	0.9	0.1	Warfare	0.213
29	0.02	50	0.9	0.5	Warfare	0.317
30	0.05	50	0.9	0.5	Warfare	−0.048
31	0.02	100	0.9	0.5	Warfare	0.386
32	0.05	100	0.9	0.5	Warfare	−1.153
33	0.02	50	0.9	0.9	Warfare	−0.005
34	0.05	50	0.9	0.9	Warfare	0.059
35	0.02	100	0.9	0.9	Warfare	−0.161
36	0.05	100	0.9	0.9	Warfare	−0.646
37	0.05	50	0.9	0.5	Warfare	0.407
38	–	50	–	–	Groups	
39	–	50	–	–	Groups	
40	–	100	–	–	Groups	
41	–	–	–	–	Economic	
42	–	–	–	–	Economic	

values of 0.1, 0.2, and 0.3, while Steponaitis derived values of 0.16–0.22 from empirical data.

Groups also call on their directly dominant and subordinate groups (but not groups from more distant portions of the complex group) for help in both attacking other groups and in defense. As complex groups are likely to have more fighters than groups that are not in complex hierarchies, being in a complex group is beneficial because more warriors leads to a greater chance of success. When fatalities occur, males are removed randomly from among all households within groups participating in the fight.

WARFARE: STOCHASTIC LANCHESTER LAWS

The models of group formation, tribute, and fighting we have described require a relevant model for the mechanics of ancient

TABLE 3 Static parameters in this sweep. All other parameters set to those used in run 230 in Kohler and Varien (2012).

Parameter	Value	Description
HUNT _ RADIUS	20	Radius for hunting (in cells; 20 cells = 4 km)
PROTEIN _ PENALTY	1	Removal of STATE _ GOOD bonus if protein needs not met (reversion to rates in life table)
NEED _ MEAT	0	Agents can move to a cell even if they cannot get enough meat via hunting
STATE _ GOOD	0.1	When an agent is good, increments birthrate by 10%, and decrements death by 10%
DOMESTICATION	TRUE	Agents can domesticate turkey
ALLIANCES	FALSE	Will groups track daughter groups and not attack them
COOP	TRUE	Agents engage in GRN and BRN exchange networks
GROUP _ BENEFIT GROWTH _ RATE	2	Growth rate for benefits as group size increases
B _ BENEFIT	73	Maximum benefit produced by contributing to the public good
C _ COST	37	Maximum cost of contributing to the public good
S _ SANCTION	56	Cost imposed on defectors. Same cost for taxation and public good defectors
CM _ MONITOR COST	4	Cost of monitoring one group member
CS _ SANCTION COST	11	Cost of sanctioning one individual, tax or public good

warfare to produce accurate probabilities of success for the aggressing or defending groups. The questions of how wars are fought and battle outcomes predicted have received ample attention elsewhere (e.g., Kress and Talmor 1999). Here we employ a set of models developed by Frederick Lanchester (Adams et al. 2003; Artelli and Deckro 2009; Kress and Talmor 1999; Lanchester 1916). Lanchester, an engineer in the British army, developed these equations to determine outcomes of air battles during World War I (Lanchester 1916) but also sought a more general description of two primary classes of warfare: "ancient" and "modern." In ancient warfare, battles were fought primarily in one-on-one duels with similar technologies (Lanchester's linear law) while in modern warfare, fighters from one team may have superior weaponry resulting in one side winning easily (Lanchester's square law). Lanchester initially derived sets of differential equations describing rates of attrition from each group under each class of warfare. These equations—now called the deterministic Lanchester laws— showed that, given equal skill of individual fighters, the larger team should win any given battle (Kress and Talmor [1999] provide a mathematical overview). These equations provide a useful means for simulating casualties in models of conflict (see, e.g., Turchin and Gavrilets 2009). However, the deterministic Lanchester laws present a problem, as intuitively we know that a smaller group must have *some* chance of winning a battle, and that its chances of winning are enhanced as the size of their forces approaches that of their enemy.

~159~

Therefore we employ probabilistic modifications of Lanchester's linear law—the stochastic Lanchester linear law—to derive the probability that a given battle will be won by a given group, following the description and formula presented by Kress and Talmor (1999). Imagine two groups (m and n) arrive at a duel-style battle (an "ancient" battle in Lanchester's estimation). Fighters on both teams possess a certain level of skill (α_m, α_n), such that a fighter with twice as much skill as its opponent will have twice the chance of winning a duel than if they were evenly matched. Each team also has an acceptable level of *attrition* (m_0, n_0), or number of casualties they are willing to endure before ceding the battle. At any given point in the battle, the number of concurrent duels in progress is equal to the minimum of the number of surviving fighters on each side. Duels take place between individuals with outcomes dependent on relative fighting skill. A new opponent from the opposing team soon thereafter meets the winner of each duel, if one is available. Fighting continues until the team with a lower attrition threshold reaches its attrition level. Thus, the probability of m winning a battle (P_m) is a function of each team's attrition thresholds and the relative strength of the fighters—the probability that team n will reach its attrition threshold before team m.

Formally, the probability that team m will win a battle may be represented as:

$$P_m = \left(\frac{1}{\alpha+1}\right)^i \sum_{i=0}^{m_0-1} \binom{n_0-1+i}{n_0-1} \cdot \left(\frac{\alpha}{\alpha+1}\right)^i$$

where $\alpha = \frac{\alpha_n}{\alpha_m}$, and the rest of the variables are as above. Clearly, $P_n = 1 - P_m$. In all the simulations reported here, we assume the skill of the fighters to be even ($\alpha = 1$) and that battles will be fought until annihilation (i.e., m_0 and n_0 are equal to the sizes of groups m and n, respectively).

Of course, it should be noted that in non-state warfare, fighting usually ceased once a group suffered a relatively small number of fatalities (Keeley 1996:91). According to Keeley, "given a high frequency of warfare ... no small group could afford to accept losses in battle exceeding 2 percent" (1996:91). Here we examine the impact of different fatality rates on our simulated populations by defining a parameter s (0.02|0.05, Table 2) to represent the acceptable proportion of fatalities to the total expected in a war of attrition (i.e., a proportion of the size of the smaller group, or sf_{mn} as above). An alternative approach might be to explore different attrition thresholds for each group, perhaps as a proportion of population, or even to "evolve" attrition threshold preferences given group experiences.

Results

To examine the effects of these specifications on long-run outcomes, we ran a sweep defined by the parameters in Tables 1–3,
searching the small space of possibilities defined by the changing
parameter values in Table 2. Where applicable, we contrast three
kinds of runs: those with territorial groups engaging in merging
and fighting; those with territorial groups but no merging or
fighting; and those with no group structure, merging, or fighting.
The runs with no group structure, merging, or fighting instantiate "Village" as described by Kohler (2012c); the other two run   ~161~
types add dynamics described here for the first time. All simulation output, as well as videos of the tribute structure, group size,
and group-type dynamics for each run are available at https://doi.
org/10.5281/zenodo.893128.

POPULATION SIZE

Figure 1 shows that the base autonomous-household-ecology
model (Village) generates fewer households through time on
average than do the other run types. The lack of constraints on
movement enjoyed by Village households is more than balanced by
the benefits received from playing the public goods game in the
other two run types. The "Groups Only" models produce the most
households because these benefits are not partially undone by mortality from warfare. (As an aside, it is likely that warfare reduces
population more in our model than it would in real populations,
since it creates a sex imbalance [only males die in warfare] that is
not compensated by polygyny, as it might be in reality.)[5]

---

[5] According to the VEPI population reconstruction (Varien et al. 2007), the area simulated here reached its maximum population in the mid-1200s AD, with some 3,200
households (~16,000 people). This peak is matched well by the "Groups Only" simulations (Fig. 1), though the other two groups of simulations underestimate this peak.
Schwindt et al. (2016) report the final VEP population estimates for the VEPIIN
study area, which at 4,600 km² is about 2.55 times larger than the VEPI study area.
Unfortunately, it is difficult to reconstruct just the VEPI population in these newer
results, so here we retain the comparison with the Varien et al. population estimates.

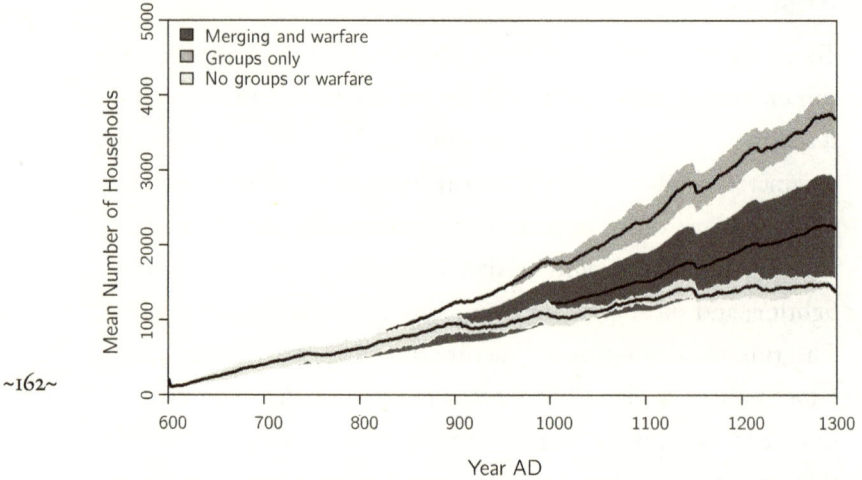

FIGURE 1 Mean number of households by run type through time. Shaded areas are one standard deviation from the mean.

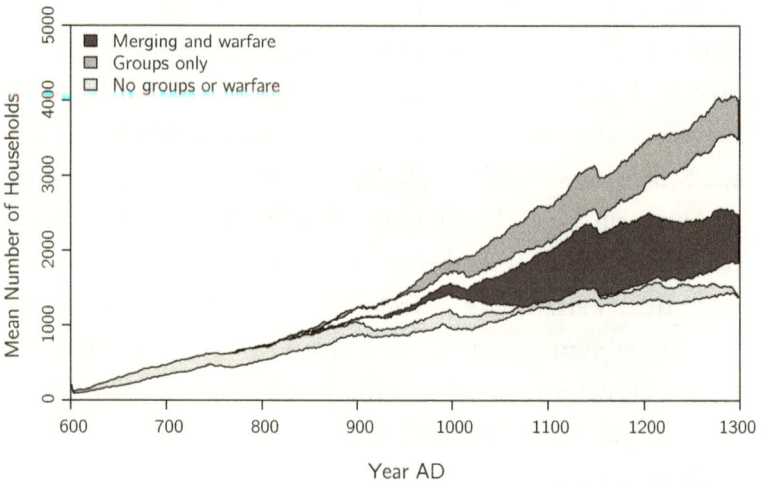

FIGURE 2 Path dependence in population size through time by run type. Each shaded area shows the difference in number of simulated households between two runs with identical parameters but different random number streams.

Considering just the runs with fighting and merging, none of the parameters listed in Table 2 has a significant effect on numbers of households through time, although higher levels of μ (proportion of the tribute from a subordinate group passed through an intermediate group to a dominant group) and lower levels of β (tax on a subordinate group's net benefit from the public goods game) are weakly associated with higher populations. We were surprised that choice of s did not significantly affect population size. These results are likely influenced by the high path dependence that we discuss next.

~163~

PATH DEPENDENCE

In most cases we performed only one run for each combination of parameters. However, we also experimented with three runs, one for each run type, duplicating parameter combinations while using different random number streams. Total populations through time for these duplicate runs are shown by run type in Figure 2, and the difference between the two duplicates is shaded in each case.

By far the least path dependence is found in the base autonomous-household-ecology model. These two runs do not diverge noticeably through time. Much more path dependence is visible in the two groups-only runs, with even more produced by the duplicated runs with both groups and warfare/merging. Variability between duplicate runs of both types increases markedly around AD 1000. We can infer that around this time households become numerous enough that the processes involving territoriality and merging/warfare introduced in these models begin to have a marked effect.

This result has two implications. First, with respect to our methods, it suggests that we will need to perform many simulations for each combination of parameters to be able to differentiate the effects of parameter choices and the effects of path dependence:

our conclusions here with respect to the effects of parameter choice must be regarded as tentative and exploratory. Second, as we will briefly argue below, these results have ramifications for our understanding of the relative importance of history and process in the analysis of historical systems, and how we approach this issue.

LINEAGE SURVIVAL THROUGH TIME

Not surprisingly, the three runs with groups but no fighting or merging tend to have a higher number of surviving lineages ($\bar{x}$ =32, σ = 1) than do the 37 runs with groups that fight and merge ($\bar{x}$ = 26.3, σ = 3.7). Considering just the runs with fighting and merging, lower values for GROUP_SIZE (50 vs. 100) significantly increase the number of surviving lineages (p = 0.02), perhaps because the greater number of groups that bud off when the span of control parameter is lower allows lineages to spread and diversify their spatial holdings. Lower values for s are weakly related to increasing the number of surviving lineages, presumably since these lower values decrease the possibility for extinction via warfare.

GROUP TYPES THROUGH TIME

Figure 3 shows the number of (simple) groups with hierarchical vs. nonhierarchical preferences through time. Not surprisingly, the three runs with groups but no fighting or merging produce far more groups by the end of the simulation ($\bar{x}$ = 252, σ = 47.8) than do the 37 runs with groups, fighting, and merging ($\bar{x}$ = 150.5, σ = 57.4). The proportion of hierarchical groups is similar for groups with no fighting or merging ($\bar{x}$ = 0.39, σ = 0.1) and for groups with fighting and merging ($\bar{x}$ = 0.36, σ = 0.1).

Complex groups can be produced only with fighting and merging. By the end of the simulation, the 37 runs with fighting and merging have an average of only 2.1 complex groups each (σ = 0.9). None of the parameters varied here has a significant effect on this

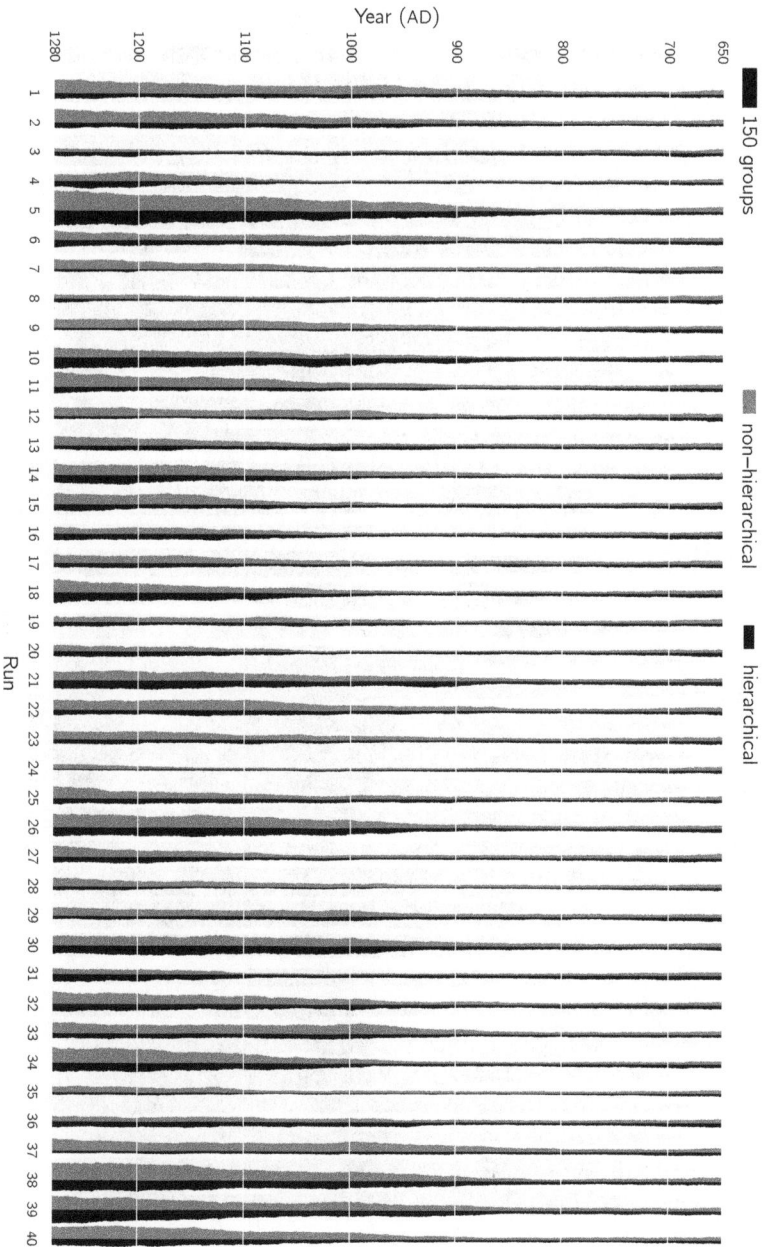

FIGURE 3 Number of hierarchical vs. nonhierarchical groups through time, per run.

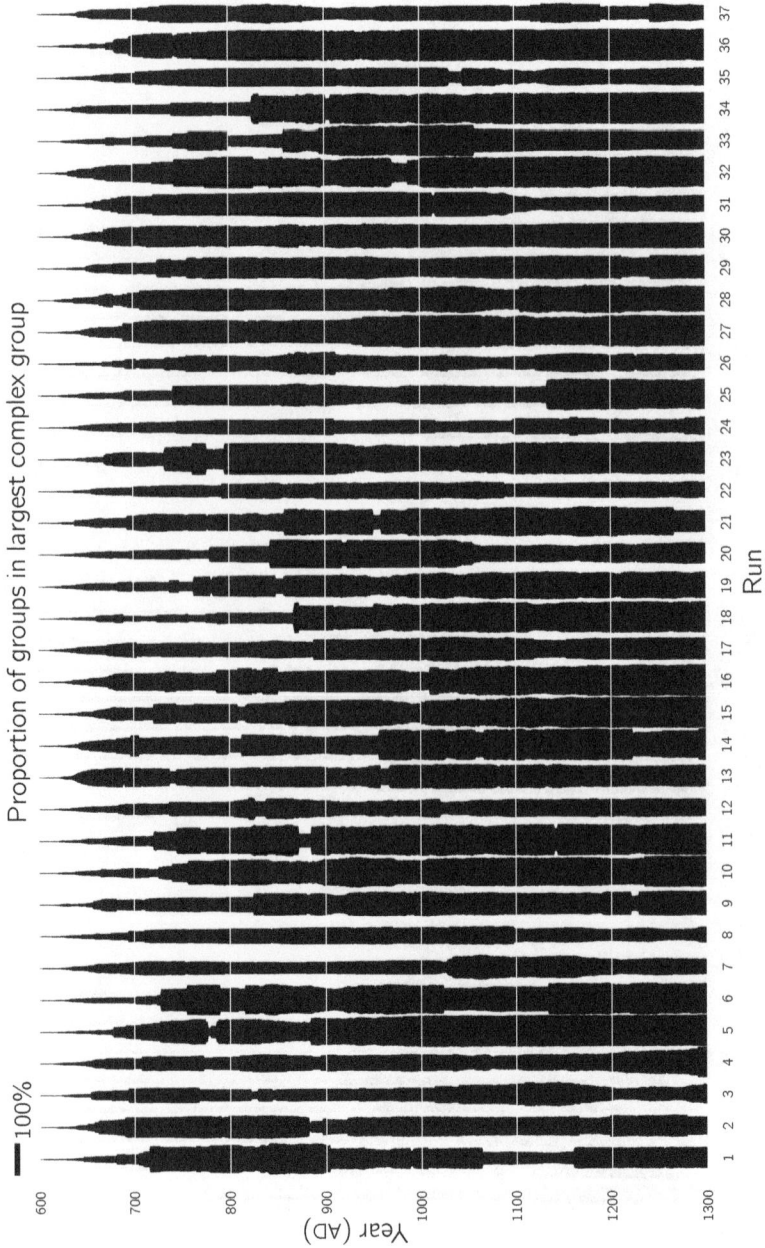

FIGURE 4 Percent of groups in the largest complex group. Wider bars indicate a greater percent in the largest group. In many runs, nearly 100 percent of groups are in the same complex group.

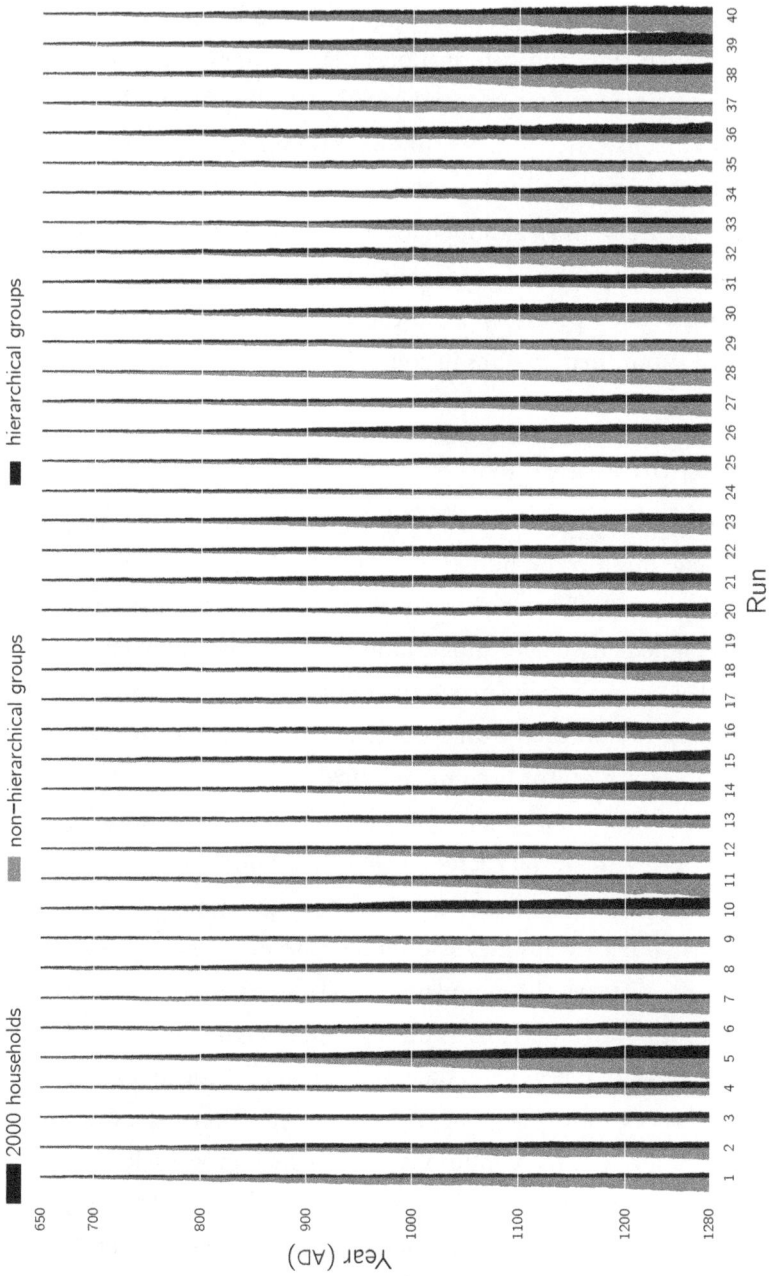

FIGURE 5 Number of households in hierarchical vs. nonhierarchical groups through time, by run.

outcome. At year 1299 (the end of the simulation) the average number of simple groups in each complex group is 117.8 ($\sigma = 64.8$). Figure 4 shows the proportion of groups in the largest complex group through time and demonstrates that this measure of concentration can wax and wane over the course of a simulation. None of the parameters varied in these runs has a significant effect on the proportion of simple groups in the largest complex group in year 1299 ($\bar{x} = 0.75$, $\sigma = 0.19$), although lower values of β (the tax on the net benefit from the public goods game) are weakly associated with higher proportions, probably because lower values of β increase the survival of subordinate groups in complex groups.

~168~

EFFECTS OF WARFARE AND MERGING ON AGENT TYPES

Compared with the results in Kohler et al. (2012), in most runs a surprising number of households end up in nonhierarchical groups (Figure 5). This is partially due to the group fissioning dynamic implemented here. Single households on the periphery of a group that has reached its maximum size will "bud off" from the parent group to start their own groups. These new groups are very often nonhierarchical (or become so quickly), and in many of the runs reported here these small groups proliferate on the landscape and rarely grow to be very large as they are almost immediately coerced into merging and paying tribute to larger groups around them. These small groups may also simply have no room to grow.

Figure 6 displays the population-level distribution of agent types through time in runs with fighting/merging (top) and without (bottom). Once again, there is surprisingly little difference between the two run types, suggesting that this model does not adequately represent the conditions under which group selection for prosociality is expected; nonhierarchical (i.e., noncooperative) types are also surprisingly numerous in almost all runs, a result substantially different from that of Kohler et al. (2012). A key

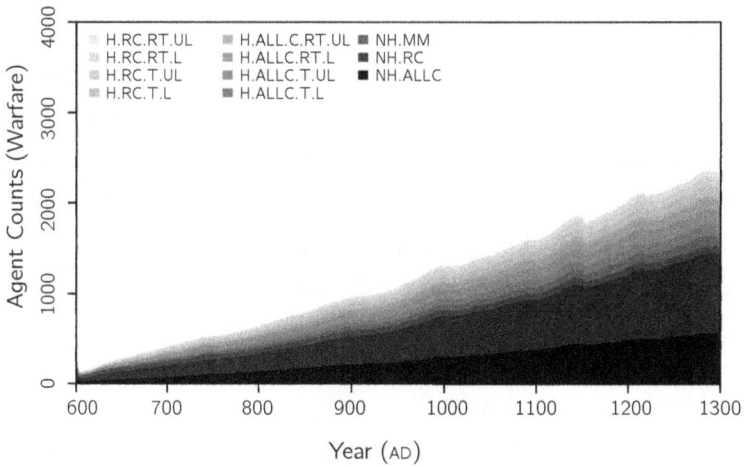

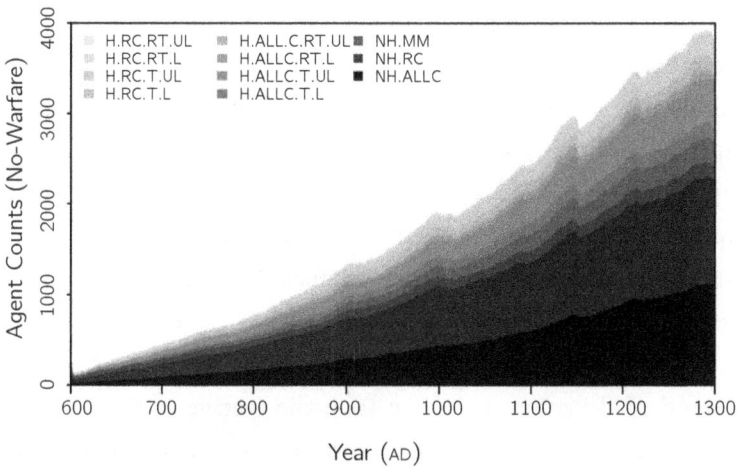

FIGURE 6 Average counts of agent types through time in runs with warfare and merging (top), and with groups but without warfare or merging (bottom). See Table 1 and Kohler et al. (2012) for definition of agent types.

difference is that here the type of group (hierarchical vs. nonhier-archical) is determined by majority rule. This allows minority pref-erences to be masked from selection. Also, because these groups are spatially constrained, they are somewhat insulated from infor-mation about the success of other strategies that might persuade them to change their minds—most nearby agents are in their own group and thus will be performing equally as well or as poorly as themselves, giving little cause to change their preference via social learning.

~170~

REALISM (VALIDATION)

It is premature to fully evaluate the goodness of fit between these simulations and their reference context at this exploratory stage, but for illustrative purposes, in the rightmost column of Table 2, we put a measure of similarity between the warfare histories and population-size histories of those runs with fighting and merging, and the reference context. For each series (violence and demog-raphy), we first took the mean across each of 14 periods for which we have accurate reference data (derived from Varien et al. 2007 and Kohler et al. 2014), and then calculated the Euclidean distance between each simulated run and the reference. We standardized each series of distances independently to have a mean of 0 and a standard deviation of 1, then took the average of the standard distances, so that similarity in the time series of population and warfare are weighted equally. To reflect similarity (as opposed to dissimilarity indicated by Euclidean distance), we negated each mean standardized distance.

None of the parameters is significantly associated with this measure of fit, though there is a very weak tendency ($p = 0.48$) for the higher level of s to be associated with better fits. The best-fit-ting run, 22, was produced by setting $s = 0.05$, GROUP_SIZE = 50, $\mu = 0.5$, and $\beta = 0.9$ (Table 2). The conflict series generated by run

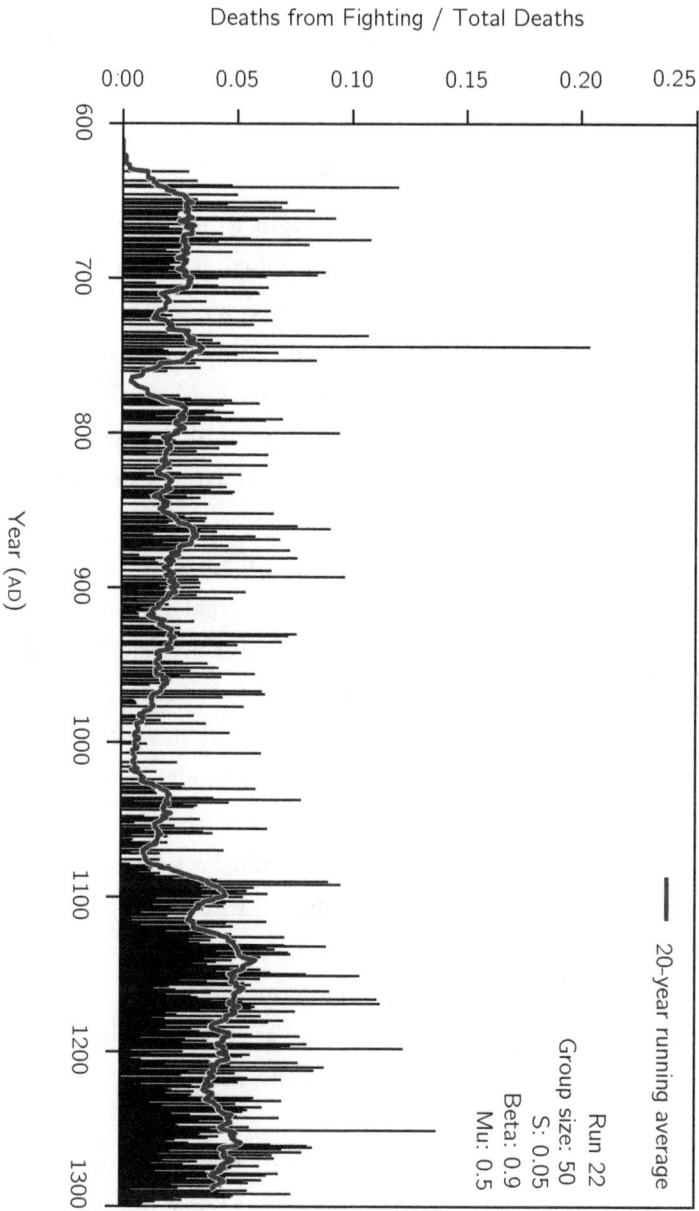

FIGURE 7 Deaths from conflict through time as a proportion of all deaths in run 22.

22 is shown in Figure 7. We emphasize, though, that another run with the same parameters but a different random number series would generate a sequence that is somewhat and perhaps substantially different.

Discussion and Conclusions

This chapter illustrates how we can begin to move beyond the verbal models that have dominated archaeological discourse on the processes by which sociopolitical scale increases—with their convenient ambiguity—to proof-of-concept computational models that unambiguously illustrate the consequences of specific model conditions and parameter values for sociopolitical change through time. Simulating these models shows what large-scale patterns emerge from clearly specified small-scale processes. We do not have to ignore micro processes to study macro outcomes, and indeed we must not. At this point, the proliferation of verbal models for processes such as polity formation no longer moves the field forward.

~172~

It is interesting to note that if the model *does* reflect sociopolitical processes approximately correctly, and if we are willing to consider the somewhat loose webs of dependencies and taxation flow we model across groups as forming polities, it is plausible to conclude that the entire VEP I area could have consisted of a single polity by the latter portions of the sequence. This possibility has also been suggested by the surprising cessation of violence as reconstructed from trauma to human bone in the late AD 1100s and early to mid-1200s (e.g., Kohler and Varien 2010). The present model suggests that political entities of this scale are indeed plausible for this period.

As with any model, we should also be careful to avoid misplaced concreteness in our interpretations. Some southwestern archaeologists who might be skeptical of "polities" in this record might be willing to entertain the possibility that what we have modeled is the emergence of networks of ceremonial dependencies and obligations, for example centered on great kivas. Ceremonial practices and obligations in these (and many other midrange) societies do seem to entail what might be considered political relations (Hooper 2012), and what we have called "leaders" here can possibly be conceptualized as leadership offices variably including priests, clowns, and other "officials." To explore this interpretation of the model we need to analyze the empirical record through time to determine the number and spatial distribution of great kivas (for example), their size hierarchy and relation to population aggregates, and the prehistory of sodalities and religious offices (see Ware [2014] for a good start).

~173~

For either the ceremonial or the political interpretation of this model, we also need to characterize the quantitative structure of the hierarchical branching networks, or "Horton orders," describing regularities in the scaling relations moving between individuals, households, extended households, roomblocks, villages of multiple roomblocks, groups of villages, and perhaps higher orders. This is feasible for areas such as Mesa Verde National Park, where we have virtually complete survey. This exercise would assist on two fronts, since it should help estimate appropriate measures of span of control for the model, and should help assess the realism of the *other* processes assumed by the model, once those estimates are correctly specified in the model. Examples of these sorts of analysis can be found in Grove (2011), Hamilton et al. (2007), and Rodriguez-Iturbe and Rinaldo (1997).

More generally, it is intriguing to consider the contributions of various processes and constraints to the high degrees of path dependence in the "histories" simulated here. The base autonomous-household-ecology model exhibits little path dependence (Kohler 2012c:71 and above). The addition of group-level territoriality considerably increases path dependence, since it

THE EMERGENCE OF PREMODERN STATES

introduces significant constraints on household movement that
depend on who previously controlled a particular patch of land,
and that prevent households from achieving an ideal free distribution. The addition of conflict and merging introduces a number
of additional probabilistic processes that deeply affect subsequent
sizes of groups, their locations, and the prominence and timing of
conflict. (Modeling revenge as an additional motive for conflict,
with the historical signal it perpetuates, would introduce even
more path dependence.)

~174~   A core ambition for historical social scientists is to weigh
the relative importance of history (or contingency) and process.
Modeling appears to be the only rigorous way to eventually move
beyond vacuous statements such as "history matters" to study the
precise ways in which history matters, and how much. Our results
suggest that the one-off history of sociopolitical complexity we
see in any specific sequence may indeed be exceptional, and if the
"tape of life" were rewound, it would not create the same record
twice. This suggests limits on our ability to retrodict (or explain)
outcomes from analysis of the processes that affect structure, and
suggests that apparently random factors early in a sequence cannot
be ignored. This can cause systems to become "locked in" to local
basins of attraction, despite conditions that would otherwise favor
sociopolitical change (Arthur 1994; Hegmon 2017; Pierson 2000).

Another question raised by our approach is under what conditions the complex groups as modeled here would tend to become
chiefdoms. Are complex groups a temporary halfway house
between tribes and chiefdoms, or are they a relatively stable organizational system that we ought to be looking for in other areas?
They bear some resemblance to the "intergroup collectivity"
described by Newman (1957; Johnson and Earle 1987:165–171)
for the Northwest Coast, except that in our model the hierarchical
groups headed by "big men" are explicitly ranked relative to each

other if they are in the same complex group. Our intuition is that rather small changes in the model—for example, allowing leaders in groups at the top of complex groups to accumulate wealth and use that to manipulate labor and obligations—would generate a system recognizably similar to a chiefdom, and perhaps such changes would result in structures more reflective of the political reality in the VEP I area during the Chaco hegemony (though the organization of that system remains controversial). The path to modeling processes of sociopolitical change in human political systems becomes increasingly clear. ⚘   ~175~

ACKNOWLEDGMENTS

The Village Ecodynamics Project was supported by the National Science Foundation (DEB-0816400 to Kohler, Allen, Kobti, and Varien). The research reported here was also made possible by support from the John Templeton Foundation ("The Principles of Complexity: Revealing the Hidden Sources of Order among the Prodigies of Nature and Culture" to the Santa Fe Institute, Grant No. 15705). We thank Jerry Sabloff for including us in this research.

REFERENCES CITED

Adams, Eldridge S., and Michael Mesterton-Gibbons
    2003   Lanchester's Attrition Models and Fights Among Social
    Animals. *Behavioral Ecology* 14(5):719–723.

Artelli, Michael, and Richard Deckro
    2009   Modeling the Lanchester Laws with System Dynamics.
    Pentagon Interior Papers. http://www.scs.org/pubs/jdms/vol5num1/
    Artelli.pdf.

Arthur, W. Brian
    1994   *Increasing Returns and Path Dependence in the Economy*.
    University of Michigan Press, Ann Arbor.

Bettinger, Robert L.
2009 Macroevolutionary Theory and Archaeology: Is There a Big Picture? In *Macroevolution in Human Prehistory: Evolutionary Theory and Processual Archaeology*, edited by Anna Marie Prentiss, Ian Kuijt, and James C. Chatters, pp. 275–295. Springer, New York.

Boone, James L.
1992 Competition, Conflict, and the Development of Hierarchies. In *Evolutionary Ecology and Human Behavior*, edited by Eric A. Smith and Bruce Winterhalder, pp. 301–337. Aldine de Gruyter, New York.

Bowles, Samuel
2009 Did Warfare Among Ancestral Hunter-Gatherers Affect the Evolution of Human Social Behaviors? *Science* 324(5932):1293–1298.

Bowles, Samuel, and Herbert Gintis
2011 *A Cooperative Species: Human Reciprocity and Its Evolution*. Princeton University Press, Princeton, New Jersey.

Boyd, Robert, and Peter J. Richerson
1985 *Culture and the Evolutionary Process*. University of Chicago Press, Chicago.
1992 Punishment Allows the Evolution of Cooperation (or Anything Else) in Sizable Groups. *Ethology and Sociobiology* 13:171–195.

Braun, David P.
1990 Selection and Evolution in Nonhierarchical Organization. In *The Evolution of Political Systems: Sociopolitics in Small-Scale Sedentary Societies*, edited by Steadman Upham, pp. 62–86. Cambridge University Press, Cambridge.

Capraro, Valerio
2013 A Model of Human Cooperation in Social Dilemmas. *PLoS ONE* 8(8):e72427. DOI:10.1371/journal.pone.0072427.

Carballo, David M., Paul Roscoe, and Gary M. Feinman
2014 Cooperation and Collective Action in the Evolution of Complex Societies. *Journal of Archaeological Method and Theory* 21:98–133.

Carneiro, Robert L.
1970 A Theory of the Origin of the State. *Science* 169:733–738.

Clark, John E.
2002 Comment on Vines of Complexity: Egalitarian Structures and the Institutionalization of Inequality among the Enga. *Current Anthropology* 43(2):255–256.

Crabtree, Stefani A.
   2015   Inferring Ancestral Pueblo Social Networks from
   Simulation in the Central Mesa Verde. *Journal of Archaeological
   Method and Theory*. DOI:10.1007/s10816-014-9233-8.

Crabtree, Stefani A., R. Kyle Bocinsky, Paul L. Hooper, Susan C. Ryan,
   and Timothy A. Kohler
   2017   How to Make a Polity (in the Central Mesa Verde Region).
   American Antiquity 82(1):71–95. DOI:10.1017/aaq.2016.18.

D'Azevedo, Warren L. (editor)
   1986   *Handbook of North American Indians: Great Basin.*
   Smithsonian Institution, Washington, DC.

Drennan, Robert D., and Christian E. Peterson   ~177~
   2012   Challenges for Comparative Study of Early Complex
   Societies. In *The Comparative Archaeology of Complex Societies*,
   edited by Michael E. Smith, pp. 62–87. Cambridge University Press,
   Cambridge.

Ember, Melvin, and Carol R. Ember
   1971   The Conditions Favoring Matrilocal versus Patrilocal
   Residence. *American Anthropologist* 73(3):571–594.

Feinman, Gary M.
   1995   The Emergence of Inequality: A Focus on Strategies and
   Processes. In *Foundations of Social Inequality*, edited by T. Douglas
   Price and Gary M. Feinman, pp. 255–279. Plenum, New York.

Flack, J. C., D. Erwin, T. Elliot, and D. C. Krakauer
   2013   Timescales, Symmetry, and Uncertainty Reduction in the
   Origins of Hierarchy in Biological Systems. In *Evolution, Cooperation
   and Complexity*, edited by K. Sterelny, Richard Joyce, Brett Calcott,
   and Ben Fraser, pp. 45–74. MIT Press, Cambridge, Massachusetts.

Flannery Kent, and Joyce Marcus
   2012   *The Creation of Inequality: How Our Prehistoric Ancestors
   Set the Stage for Monarchy, Slavery, and Empire.* Harvard University
   Press, Cambridge, Massachusetts.

Fried, M. H.
   1967   *The Evolution of Political Society: An Essay in Political
   Anthropology.* Random House, New York.

Gavrilets, Sergey, David G. Anderson, and Peter Turchin
   2010 Cycling in the Complexity of Early Societies. *Cliodynamics: The
   Journal of Theoretical and Mathematical History* 1(1):58–80. http://
   escholarship.org/us/item/5536t55r.

Gibson, D. Blair
    2008   Chiefdoms and the Emergence of Private Property in Land. *Journal of Anthropological Archaeology* 27:46–62.

Gould, Stephen J., and Richard C. Lewontin
    1979   The Spandrels of San Marco and the Panglossian Paradigm. *Proceedings of the Royal Society of London B* 205(1161):581–598.

Grove, Matt
    2011   An Archaeological Signature of Multi-Level Social Systems: The Case of the Irish Bronze Age. *Journal of Anthropological Archaeology* 30:44–61.

Haas, Jonathan, Edmund J. Ladd, Jerrold E. Levy, Randall H. McGuire, and Norman Yoffee
    1994   Historical Processes in the Prehistoric Southwest. In *Understanding Complexity in the Prehistoric Southwest*, edited by G. J. Gumerman and Murray Gell-Mann, pp. 203–232. Proceedings Volume XVI, Santa Fe Institute. Addison-Wesley, Reading, Massachusetts.

Hamilton, M. J., B. T. Milne, R. S. Walker, O. Burger, and J. H. Brown
    2007   The Complex Structure of Hunter-Gatherer Social Networks. *Proceedings of the Royal Society B – Biological Sciences* 274:2195–2202.

Hegmon, Michelle
    2017   Path Dependence. In *Oxford Handbook of the Archaeology of the Southwest*, edited by Barbara J. Mills and Severin Fowles, pp. 155–166. Oxford University Press, Oxford.

Henrich, Joseph
    2004   Cultural Group Selection, Coevolutionary Processes and Large-Scale Cooperation. *Journal of Economic Behavior & Organization* 53(1):3–35.

Hobbes, Thomas
    1651[1957]   *Leviathan or the Matter, Forme and Power of a Commonwealth Ecclesiasticall and Civil*. Basil Blackwell, Oxford.

Holland, John H.
    2014   *Complexity: A Very Short Introduction*. Oxford University Press, Oxford.

Hooper, Paul L.
    2012   Modeling the Evolution of Religious Institutions. *Religion, Brain & Behavior* 2(3):209–212.

Hooper, Paul L., Kathryn Demps, Michael Gurven, Drew Gerkey, and Hillard S. Kaplan
2015   Skills, Division of Labour and Economies of Scale Among Amazonian Hunters and South Indian Honey Collectors. *Philosophical Transactions of the Royal Society B* 370:20150008.

Hooper, Paul L., Hillard S. Kaplan, and James L. Boone
2010   A Theory of Leadership in Human Cooperative Groups. *Journal of Theoretical Biology* 265:633–646.

Johnson, Allen W., and Timothy Earle
1987   *The Evolution of Human Societies: From Foraging Group to Agrarian State.* Stanford University Press, Stanford, California.

Johnson, C. David, and Timothy A. Kohler   ~179~
2012   Modeling Plant and Animal Productivity and Fuel Use. In *Emergence and Collapse of Early Villages: Models of Central Mesa Verde Archaeology*, edited by Timothy A. Kohler and Mark D. Varien, pp. 113–128. University of California, Berkeley.

Johnson, Gregory A.
1978   Information Sources and the Development of Decision-Making Organizations. In *Social Archaeology: Beyond Subsistence and Dating*, edited by Charles L. Redman et al., pp. 87–112. Academic Press, New York.
1982   Organizational Structure and Scalar Stress. In *Theory and Explanation in Archaeology*, edited by Colin Renfrew, Mark J. Rowlands, and Barbara A. Segraves, pp. 389–421. Academic Press, New York.

Ka'ai-Mahuta, Rachael Te Āwhina
2010   He kupu tuku iho mō tēnei reanga: A Critical Analysis of *Waiata* and *Haka* as Commentaries and Archives of Māori Political History. Unpublished PhD thesis. Auckland University of Technology. http://aut.researchgateway.ac.nz/bitstream/handle/10292/1023/Kaai_MahutaR.pdf.

Keeley, Lawrence H.
1996   *War Before Civilization.* Oxford University Press, Oxford.

Kobti, Ziad
2012   Simulating Household Exchange with Cultural Algorithms. In *Emergence and Collapse of Early Villages: Models of Central Mesa Verde Archaeology*, edited by Timothy A. Kohler and Mark D. Varien, pp. 165–174. University of California Press, Berkeley.

Kohler, Timothy A.
2012a   Complex Systems and Archaeology. In *Archaeological Theory Today*. 2nd ed., edited by Ian Hodder, pp. 93–123. Polity Press, Cambridge, Massachusetts.
2012b   Modeling Agricultural Variability and Farming Effort. In *Emergence and Collapse of Early Villages: Models of Central Mesa Verde Archaeology*, edited by Timothy A. Kohler and Mark D. Varien, pp. 85–112. University of California, Berkeley.
2012c   Simulation Model Overview. In *Emergence and Collapse of Early Villages: Models of Central Mesa Verde Archaeology*, edited by Timothy A. Kohler and Mark D. Varien, pp. 59–72. University of California, Berkeley.

~180~

Kohler, Timothy A., Denton Cockburn, Paul Hooper, R. Kyle Bocinsky, and Ziad Kobti
2012   The Coevolution of Group Size and Leadership: An Agent-Based Public Goods Model for Prehispanic Pueblo Societies. *Advances in Complex Systems* 15:29 pages.

Kohler, Timothy A., and Laura Ellyson
2018   In and Out of Chains? The Changing Social Contract in the Pueblo Southwest, AD 600–1300. In *Ten Thousand Years of Inequality: The Archaeology of Wealth Differences*, edited by Timothy A. Kohler and Michael E. Smith, pp. 130–154. University of Arizona Press, Tucson.

Kohler, Timothy A., and Rebecca Higgins
2016   Quantifying Household Inequality in Early Pueblo Villages. *Current Anthropology* 57(5):690–697.

Kohler, Timothy A., Scott G. Ortman, Katie E. Grundtisch, Carly M. Fitzpatrick, and Sarah M. Cole
2014   The Better Angels of Their Nature: Declining Violence Through Time among Prehispanic Farmers of the Pueblo Southwest. *American Antiquity* 79(3):444–464.

Kohler, Timothy A., and Charles Reed
2011   Explaining the Structure and Timing of Formation of Pueblo I Villages in the Northern U.S. Southwest. In *Sustainable Lifeways: Cultural Persistence in an Ever-changing Environment*, edited by Naomi F. Miller, Katherine M. Moore, and Kathleen Ryan, pp. 150–179. University of Pennsylvania Museum of Archaeology and Anthropology, Philadelphia.

Kohler, Timothy A., and Mark D. Varien
    2010   A Scale Model of Seven Hundred Years of Farming
    Settlements in Southwestern Colorado. In *Becoming Villagers:*
    Comparing Early Village Societies, edited by Matthew S. Bandy and
    Jake R. Fox, pp. 37–61. University of Arizona Press, Tucson.

Kolm, Kenneth E., and Schaun M. Smith
    2012   Modeling Paleohydrological System Structure and Function.
    In *Emergence and Collapse of Early Villages: Models of Central Mesa*
    Verde Archaeology, edited by Timothy A. Kohler and Mark D. Varien,
    pp. 73–84. University of California, Berkeley.

Kress, M., and I. Talmor
    1999   A New Look at the 3:1 Rule of Combat Through Markov   ~181~
    Stochastic Lanchester Models. *Journal of the Operational Research*
    Society 50(7):733–744.

Lanchester, F. W.
    1916   *Aircraft in Warfare*. Appleton, New York.
    1956   Mathematics in Warfare. In *The World of Mathematics*, vol.
    4, edited by J. R. Newman, pp. 2138–2157. Simon and Schuster, New
    York.

Levy, Jerrold E.
    1992   *Orayvi Revisited: Social Stratification in an "Egalitarian"*
    Society. School of American Research Press, Santa Fe, New Mexico.

Lightfoot, Kent G., and Steadman Upham
    1989   Complex Societies in the Prehistoric American Southwest:
    A Consideration of the Controversy. In *The Sociopolitical Structure*
    of Prehistoric Southwestern Societies, edited by Steadman Upham,
    Kent G. Lightfoot, and Roberta A. Jewett, pp. 3–30. Westview Press,
    Boulder, Colorado.

Mahoney, Nancy M., and John Kantner
    2000   Chacoan Archaeology and Great House Communities.
    Great House Communities Across the Chacoan Landscape, edited
    by John Kantner and Nancy M. Mahoney, pp. 1–18. University of
    Arizona Press, Tucson.

Maine, Henry S.
    1861[1931]   *Ancient Law*. Oxford University Press, London.

Manson, J. H., and R. W. Wrangham
    1991   Intergroup Aggression in Chimpanzees and Humans.
    Current Anthropology 32:369–390.

Mathew, Sarah, and Robert Boyd
    2013   The Cost of Cowardice: Punitive Sentiments Towards Free
    Riders in Turkana Raids. *Evolution and Human Behavior* 35:58–64.

Mezza-Garcia, Nathalie, Tom Froese, and Nelson Fernández
    2014   Reflections on the Complexity of Ancient Social
    Heterarchies: Toward New Models of Social Self-Organization in Pre-
    Hispanic Colombia. *Journal of Sociocybernetics* 12:3–17.

Netting, Robert McC.
    1990   Population, Permanent Agriculture, and Polities:
    Unpacking the Evolutionary Portmanteau. In *The Evolution of polit-
ical Systems: Sociopolitics in Small-Scale Sedentary Societies*, edited
    by Steadman Upham, pp. 21–61. Cambridge University Press,
    Cambridge.

Newman, P.
    1957   An Intergroup Collectivity Among the Nootka. Master's
    thesis, Department of Anthropology, University of Washington,
    Seattle.

Olson, Mancur
    1971   *The Logic of Collective Action: Public Goods and the Theory of
Groups*. Harvard University Press, Cambridge, Massachusetts.

Ortman, Scott G., Donna M. Glowacki, Mark D. Varien, and C. David
    Johnson
    2012   The Study Area and the Ancestral Pueblo Occupation.
    In *Emergence and Collapse of Early Villages*, edited by Timothy A.
    Kohler and Mark D. Varien, pp. 15–40. University of California Press,
    Berkeley.

Pierson, Paul
    2000   Increasing Returns, Path Dependence, and the Study of
    Politics. *American Political Science Review* 94:251–267.

Plog, Stephen
    1990   Agriculture, Sedentism, and Environment in the Evolution
    of Political Systems. In *The Evolution of Political Systems: Sociopolitics
in Small-Scale Sedentary Societies*, edited by Steadman Upham, pp.
    177–199. Cambridge University Press, Cambridge.

Reese, Kelsey M.
    2014   Over the Line: A Least-Cost Analysis of "Community" in
    Mesa Verde National Park. Unpublished master's thesis, Department
    of Anthropology, Washington State University, Pullman.

Rodriguez-Iturbe, I., and A. Rinaldo
    1997   *Fractal River Basins: Chance and Self-Organization*.
    Cambridge University Press, Cambridge.

Rogers, Deborah S., Omkar Deshpande, and Marcus W. Feldman
    2011   The Spread of Inequality. *PLoS ONE* 6(9):e24683.
    DOI:10.1371/journal.pone.0024683.

Rosenberg, Michael
  2009   Proximate Causation, Group Selection, and the Evolution
  of Hierarchical Human Societies: System, Process, and Pattern. In
  *Macroevolution in Human Prehistory: Evolutionary Theory and
  Processual Archaeology*, edited by Anna Marie Prentiss, Ian Kuijt, and
  James C. Chatters, pp. 23–49. Springer, New York.

Schwindt, D. M., R. K. Bocinsky, S. G. Ortman, D. M. Glowacki, M. D.
  Varien, and T. A. Kohler
  2016   The Social Consequences of Climate Change in the Central
  Mesa Verde Region. *American Antiquity* 81(1):74–96.

Service, Elman R.
  1962   *Primitive Social Organization: An Evolutionary Perspective.*
  2nd ed. Random House, New York.

Smith, Eric A., and Jung-Kyoo Choi
  2007   The Emergence of Inequality in Small-Scale Societies:
  Simple Scenarios and Agent-Based Simulations. In *The Model-Based
  Archaeology of Socionatural Systems*, edited by Timothy A. Kohler
  and Sander E. van der Leeuw, pp. 105–120. School for Advanced
  Research, Santa Fe, New Mexico.

Stanish, Charles
  2009   The Evolution of Managerial Elites in Intermediate
  Societies. In *The Evolution of Leadership: Transitions in Decision
  Making from Small-Scale to Middle-Range Societies*, edited by Kevin
  J. Vaughn, Jelmer W. Eerkens, and John Kantner, pp. 97–119. SAR
  Press, Santa Fe, New Mexico.

Steponaitis, Vincas P.
  1981   Settlement Hierarchies and Political Complexity in
  Nonmarket Societies: The Formative Period of the Valley of Mexico.
  American Anthropologist 83(2):320–363.

Turchin, Peter, and Sergey Gavrilets
  2009   Evolution of Complex Hierarchical Societies. *Social
  Evolution & History* 8(2):167–198.

Upham, Steadman
  1990   Decoupling the Processes of Political Evolution. In *The
  Evolution of Political Systems: Sociopolitics in Small-Scale Sedentary
  Societies*, edited by Steadman Upham, pp. 1–17. Cambridge University
  Press, Cambridge.

Upham, Steadman (editor)
  1990   *The Evolution of Political Systems: Sociopolitics in Small-scale
  Sedentary Societies* Cambridge University Press, Cambridge.

~183~

Varien, Mark D.
1999   *Sedentism and Mobility in a Social Landscape: Mesa Verde and Beyond.* University of Arizona Press, Tucson.

Varien, Mark D., Scott G. Ortman, Timothy A. Kohler, Donna M. Glowacki, and C. David Johnson
2007   Historical Ecology in the Mesa Verde Region: Results From the Village Project. *American Antiquity* 72:273–299.

von Rueden, Christopher, Michael Gurven, Hillard Kaplan, and Jonathan Stieglitz
2014   Leadership in an Egalitarian Society. *Human Nature* DOI:10.1007/s12110-014-9213-4.

Ware, John A.
2014   *A Pueblo Social History: Kinship, Sodality, and Community in the Northern Southwest.* SAR Press, Santa Fe, New Mexico.

Weiss, Kenneth M.
1973   *Demographic Models for Anthropology. Memoirs of the Society for American Archaeology, no. 27.* Salt Lake City, Utah.

Wiessner, Polly
2002   Vines of Complexity: Egalitarian Structures and the Institutionalization of Inequality Among the Enga. *Current Anthropology* 43(2):233–269.

Wilson, David Sloan, and Edward O. Wilson
2007   Rethinking the Theoretical Foundation of Sociobiology. *Quarterly Review of Biology* 82(4):327–348.

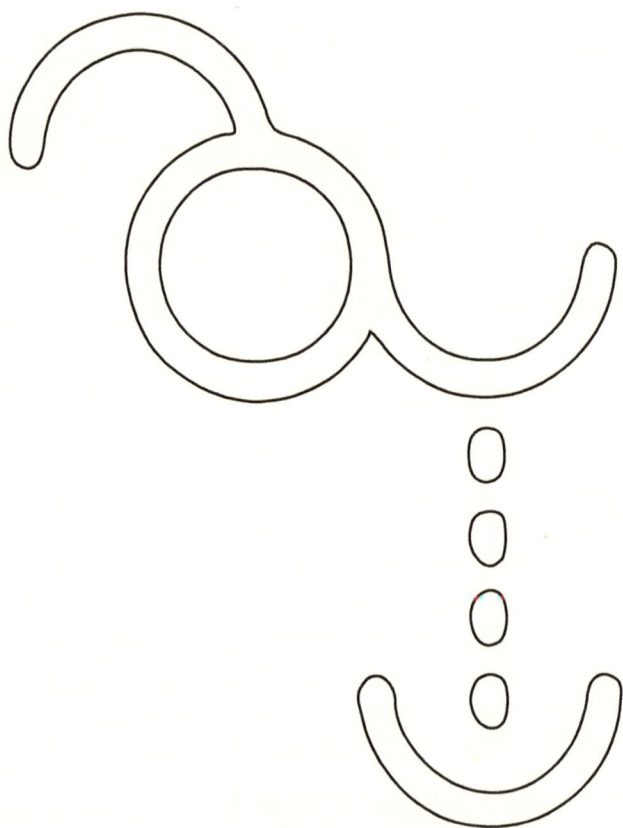

N

THE CONTOURS OF CULTURAL EVOLUTION

Scott G. Ortman, University of Colorado Boulder and Santa Fe Institute
Lily Blair, Stanford University
and Peter N. Peregrine, Lawrence University and Santa Fe Institute

It is undeniable that human societies have tended to grow in scale and complexity over time, and especially over the past 12,000 years. Understanding this general phenomenon has been and continues to be one of the central pursuits of archaeology, but understanding of the mechanisms behind this basic pattern remains elusive. One reason for this is the rapid development of archaeological method and theory, which has made it possible to reconstruct past social dynamics in great detail. As a result, the unique aspects of each society come to the foreground, encouraging a view of social evolution as myriad sequences of historically contingent events. As a result, generalizations concerning universal patterns in social evolution can seem counterproductive: for any generalization one researcher makes, another can respond with a fine-grained analysis of a specific society which shows how specific and unique details governed the sequence of change in that case (e.g., Pauketat 2007; Smith 2003; Yoffee 2005). Such studies often conclude that local details better explain the evolution of a particular society and thus undermine the validity of general principles or processes.

When this situation is encountered in other fields, a typical solution is "coarse-graining"—stepping back from the details to a level of focus where the behavior of the system becomes more amenable to generalization. So, for example, in physics, the specific trajectories of individual gas molecules in a chamber are the result of myriad contingent events, but the aggregate behavior of many gas molecules is well approximated by the ideal gas law $PV = nRT$; and in biology, specific episodes of adaptive radiation appear to be

THE EMERGENCE OF PREMODERN STATES

historically contingent (Gould 1989; Sallan and Coates 2010), but there are also general patterns in adaptive radiation that Darwinian evolution accounts for readily (Gavrilets and Losos 2009; Gavrilets and Vose 2005). In biology this scale of analysis is known as *macroevolution* or *macroecology* (Brown 1995; Sepkoski 2012). In this chapter we suggest that a similar approach is useful for building a general understanding of human social evolution. We examine this process from a high altitude, where societies still vary in their basic properties but the patterns in this variation appear more regular than they do when viewed in close detail. We acknowledge that a wealth of local detail exists and is important for understanding specific trajectories, but argue that there are still regularities amenable to generalization.

~188~

We find that a complex systems perspective, which views human societies as dynamic networks of people, energy, and information that exhibit emergent properties related to their structure and functioning, is useful for developing a general understanding of the changes that have occurred in human societies over the past 12,000 years. In this chapter we suggest that it is useful to conceive of cultural evolution as a macroevolutionary process driven by: (1) intrinsic economies of scale in social organization; (2) elaboration of the division and coordination of labor enabled by these economies of scale; and (3) increases in the ability of social networks to capture and distribute energy. We also view cultural macroevolution as a cumulative process in which innovation derives from the recombination of existing elements into new structures, which then become elements for further combinations, and so forth (Arthur 2009; Gell-Mann 2011; Peregrine et al. 2004).

Our macroevolutionary perspective addresses one of the fundamental shortcomings of many previous studies of social evolution in archaeology. Although the scale and complexity of human societies have generally increased over time, the process has been uneven, with many local reversals or periods of stasis. Yet researchers seeking to develop general theory have typically focused on cases of primary state formation (Mesopotamia, Egypt, China, Peru, Mesoamerica,

Although the scale and complexity of human societies have generally increased over time, the process has been uneven, with many local reversals or periods of stasis.

the Indus Valley, and Hawai'i) while neglecting cases where com- ~189~
plexity has *not* accumulated or has done so in a different way. Even comparative studies typically compare archaeological cases where "complex" societies first emerged (Adams 1966; Blanton and Fargher 2008; Trigger 2003; Yoffee 2005) or arrange ethnographic societies along an implied developmental pathway (Flannery and Marcus 2012; Johnson and Earle 1987). Such studies typically ask why certain societies became complex but rarely ask why others did not. Most theories concerning social evolution make at least implicit predictions about this broader range of cases, but such predictions are rarely checked systematically. Thus, a macroevolutionary approach, which considers patterns of change in all human societies, is a necessary corrective.

A Global Perspective on Social Complexity

Our exploration of cultural macroevolution utilizes a database of basic information on archaeological traditions from across the globe that we compiled for this project. The backbone of the database is the *Atlas of Cultural Evolution* (ACE), a database modeled after the Human Relations Area Files compiled by Peregrine (2003) using information contained in Peregrine and Ember's (2001–2002) *Encyclopedia of Prehistory*. Peregrine and Ember compiled data for 289 archaeological traditions—which they define as a group of populations sharing similar subsistence practices, technology, and forms of sociopolitical organization across a contiguous area and over a long period—dating from the dawn of modern humans to the onset of written history in various world areas. These are not equivalent to ethnographic cultures, but there is a sense in which these traditions represent distinct adaptations to particular socio-natural contexts. It is also important to note that major cultural periods in well-studied areas are also divided into separate traditions; so Mesoamerica, for example, is represented by 13 traditions (Highlands Archaic, Lowlands Archaic, Highland Early Preclassic, Highland Late Preclassic, Maya Preclassic, Olmec, Central Mexico Classic, Southern Highland Classic, Gulf Coast Classic, Classic Maya, Central Mexico Postclassic, Southern Highland Postclassic, and Postclassic Maya), the US Southwest by 10 (Middle Desert Archaic, Early Hohokam, Late Hohokam, Patayan, Early Mogollon, Late Mogollon, Basketmaker, Early Anasazi, Late Anasazi, and Fremont), and so forth. The ACE contains information on the location, duration, and ancestor–descendant relationships of each tradition, as well as coded information on the following variables: writing and record keeping, fixity of residence, degree of agricultural dependence, degree of urbanization, technical knowledge of materials, forms of land transport, forms of currency, population density, scale of political integration, and degree of social stratification. Each tradition is coded on a three-point scale for each of these 10 variables, and these scores can be summed to produce an overall complexity score for each tradition ranging from 10 to 30. These 10 ordinal-scale variables can also be transformed into a list of 30 presence/absence attributes, and Peregrine and others (2004) have shown that these attributes can be arranged in an implicational or Guttman scale in the same way that cross-cultural ethnographic data have been arranged in previous studies (Carneiro 1962). These data thus replicate the results of cross-cultural ethnographic studies in suggesting that social evolution has a cumulative character.

For this study we added a variety of additional data for each tradition: the population and area of the largest settlement (from the literature), the time elapsed from the onset of agricultural dependence (calculated from the ACE data), the surface area encompassed by remains of each tradition (from maps in the *Encyclopedia of Prehistory*), and statistical summaries of the net primary productivity of these areas (a measure of the rate of carbon fixation by plants, from the Atlas of the Biosphere (see https://nelson.wisc. edu/sage/data-and-models/atlas/). We also added two biological measures: (1) a health index derived from studies of human skeletal remains for New World traditions (Steckel and Rose 2002); and (2) the fraction of juveniles (5- to 15-year-olds) among human skeletal remains from that tradition (Boquet-Appel 2002; Boquet-Appel and Bar-Yosef 2008; Boquet-Appel and Naji 2006; Kohler and Reese 2014). The latter has been found to be a good proxy for the crude birth rate (Boquet-Appel 2002; Kohler and Reese 2014) and thus provides some indication of the overall demographic potential of an adaptation. Finally, we coded a variety of attributes related to technology, the division of labor, the provision of public goods, and information processing that are observable from the archaeological record for a sample of North American and Mesoamerican traditions based on information in the *Encyclopedia of Prehistory*. The list of attributes coded is given in Table 1, and a summary of the resulting database is presented in the appendix to this volume (see page 308). Many studies of social evolution have utilized nominal-scale data from cross-cultural ethnographic samples (Carneiro 1967, 2000; Chick 1997; Feinman 2011; Feinman and Neitzel 1984; Naroll 1956). The data compiled for this project allow us to examine similar questions using continuous measures from diachronic archaeological cases.

THE EMERGENCE OF PREMODERN STATES

TABLE 1 List of attributes coded for a sample of North American and Mesoamerican traditions (data are presence/absence unless otherwise noted).

Attribute	Attribute
Differentiation of dwellings	Horticulture (gardening; no fields)
Dwellings divided into "rooms"	Dry farming
Seasonal settlements (camps)	Irrigation
Permanent settlements (village, hamlet)	Plow (presence)
Settlement hierarchy	Domestication for industrial use (twine, clothing, etc.)
Cities (Presence/absence)	Domestication for drugs: tobacco, coca, chocolate, chili
Ceremonial structures	Domestication of animals: food
Public/administrative buildings	Domestication of animals: labor
Manufacturing centers	Domestication of animals: materials
From chiefdom to state organization (Range: 0–4)	Utilitarian ceramics (domestic)
Social classes (Range: 0–3)	Non-utilitarian ceramics (display or ritual)
Slave/corvée/caste	Kilns
Interpersonal conflict	Mass-produced ceramics
Personal ornaments	Metal tools
Burials/graves with offerings	Metal ornaments
Formal cemeteries	Metal status goods
Tombs/mausoleums	Metal for trade (export)
Public works: Protection from natural elements (ditches)	Cold-worked (hammered)
Ditches/moats/dams/reservoirs/drains/wells	Smelted/melted
Public works: Divine projects (altars, pyramids, temples)	Imported raw materials
Public works: Administration	Imported finished goods
City walls (for defense, etc.)	Specialized craftsmen
Public works: Defense (e.g., walls, signal towers, etc.)	Seals, labels, stamps
Public works: Territorial boundaries (walls, etc.)	Weights
Streets/roads/alleys	Documents (tablets, papyrus)
Indication of extraction (taxes, tribute)	Writing

TABLE 1 (*continued*)

Attribute	Attribute
Public works: Prestige (of ruler/ruling class)	Counters (quipu, abacus, etc.)
Organized conflict (war)	Media of exchange (e.g., cacao, coins)
Missiles (spears, bow/arrow, etc.)	Wall/cave pictures/decoration
Clubs, knives, mallets	Portable art (sculpture)
Armor	Nonportable art (permanent art, e.g., stone sculpture)

Measuring Cultural Complexity ~193~

Human societies vary along as many dimensions as one could choose to measure, but for a variety of reasons we feel the logarithm of the largest settlement population is as good a single measure of social complexity as one could hope to recover from a broad range of archaeological traditions. Although the logarithm of the largest settlement population is most directly related to the scale of social network organization, there are a number of reasons why it is also a good proxy for overall complexity. First, because the largest settlements have long attracted the most archaeological attention, the largest settlements for the majority of archaeological traditions are known and their resident populations have been estimated in some way. Second, ethnographic studies of social complexity (Carneiro 1967; Chick 1997; Naroll 1956) have found that the logarithm of the largest settlement population is correlated with multivariate indices of social complexity drawn from the standard cross-cultural sample. Third, this same relationship is apparent in a comparison of largest settlement populations with the complexity score for archaeological traditions in the ACE (Figure 1), but the former is a continuous variable that ranges over five orders of magnitude whereas the latter is an ordinal-scale variable that ranges only from 10 to 30. Finally, studies of entire settlement systems have often found that the distribution of settlement populations is

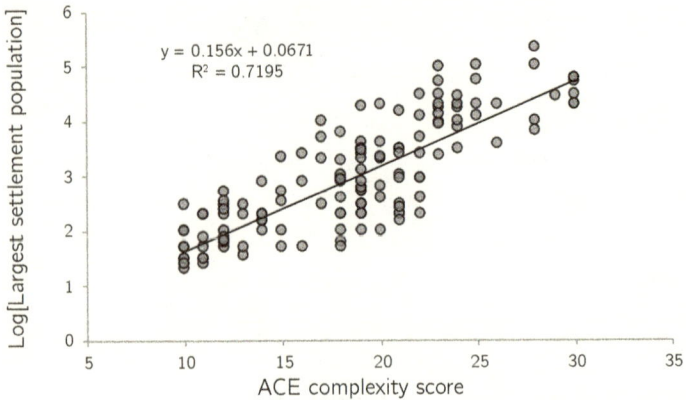

$y = 0.156x + 0.0671$
$R^2 = 0.7195$

FIGURE 1 Correlation between ACE complexity score and the log of the largest settlement population.

well approximated by Zipf's law or the rank-size rule, which states that the population size of the nth ranked settlement (by population) is equal to the population of the largest settlement divided by n (Johnson 1977, 1980, 1987). So if the population of the largest settlement is known, characteristics of the larger settlement system of which it was a part can often be inferred, even if data for the larger system are unavailable. For these reasons, we use the logarithm of the population of the largest settlement associated with each archaeological tradition as a measure of the scale of organization of that tradition, and as a proxy for its overall complexity, to assess general patterns in cultural macroevolution.

What Cultural Macroevolution Is Not

The first point we wish to make is that the data compiled for this project help define what cultural macroevolution is by showing what it is not. Figure 2, for example, demonstrates that the evolution of social complexity has not been an even process: the range of variation in the scale of human societies has grown dramatically over the past 12,000 years, but relatively small-scale archaeological traditions have persisted until recent times. Thus, the level of complexity of archaeological traditions has not been uniform across the globe at any given time. The continuous expansion of this "envelope of complexity" does suggest that there are advantages to scale, but the persistence of small-scale traditions throughout prehistory suggests that small-scale societies also work well. A good macroevolutionary theory needs to account for both aspects of this pattern.

Figure 2 also raises the possibility that there is no intrinsic biological advantage to social complexity. We explore this hypothesis further in Figure 3, which compares social complexity with measures of health and crude birth rates derived from studies of human skeletal remains. The quality of life index (QALY) integrates commonly recorded skeletal health indicators across life spans to derive a single measure of the biological quality of life (Steckel and Rose 2002); the juvenility index is the fraction of all individuals who lived to be at least five years old that died between the ages of five and 19 (Boquet-Appel and Bar-Yosef 2008; Kohler and Reese 2014). These data suggest there is little or no relationship between social complexity and group-level health or birth rates.

~195~

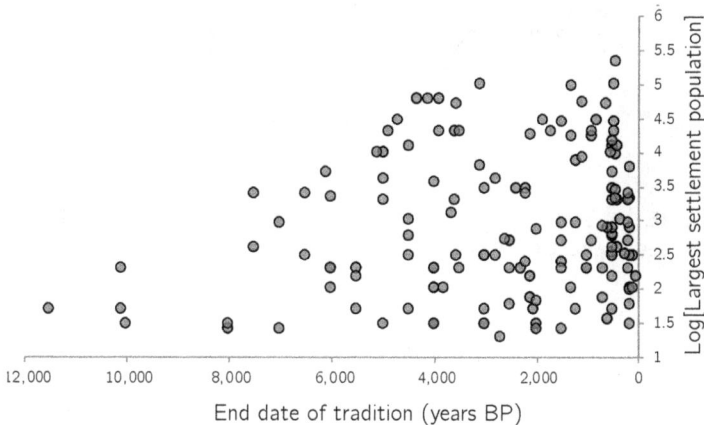

FIGURE 2 Time vs. scale of archaeological traditions.

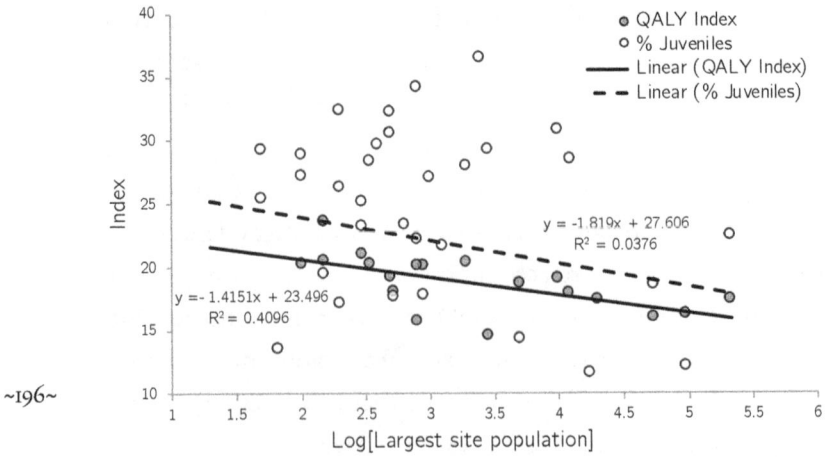

FIGURE 3 Scale vs. health and birth rate for archaeological traditions.

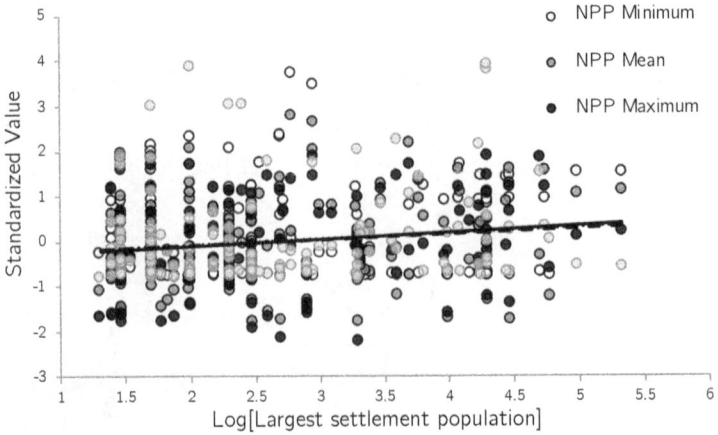

FIGURE 4 Scale vs. intrinsic productivity for archaeological traditions (see the text for regression results).

It is important to emphasize that the interpretation of paleopathology data is tricky because pathology rates in skeletal remains are
a product of both the incidence of the underlying conditions and
the rate at which people with those conditions survive to the point
that the conditions manifest in their bones (Wood et al. 1992); the
interpretation of age at death distributions also assumes a representative sample of all individuals who died at age five or older. So
the incidence of skeletal pathologies may reflect increased survival
of frail individuals as opposed to increased incidence of underlying
health stressors, and juvenility indices may reflect patterns in burial
practices as opposed to crude birth rates. Despite these caveats, the
lack of a strong relationship between health indicators and social
complexity suggests that complexity can accumulate somewhat
independently of its direct biological consequences at the group
level.

~197~

We also note that the accumulation of social complexity has
not been driven by the intrinsic productive capacity of local environments. Figure 4 compares summary measures of the net primary
productivity (NPP; kg of carbon captured per m^2 per year) across 1
degree pixels within each archaeological tradition to the scale and
complexity of that tradition. If salubrious environments were all
that it took for complexity to accumulate, one would expect the
range of complexity values to increase along with various measures of NPP. Or, if pressures toward complexity were strongest in
the most difficult environments, one would expect the opposite
pattern. Figure 4 shows that there is no association between larger-scale traditions and larger NPP measures. None of the regressions are significant at the 0.05 level, and the R^2 values are all below
0.03. Thus, although environmental variation has certainly played
a role in the character of economic systems that emerged in various
times and places, the intrinsic productivity of environments has
not been a basic driver of complexity.

Finally, social complexity does not appear to be the necessary result of any particular technological innovation. Figure 5 illustrates this point for perhaps the most important innovation in human history—the production of food from domesticated plants. This plot compares the length of time that had passed since the ancestors of an archaeological tradition became dependent upon agriculture to the complexity of that tradition. This analysis shows that complexity did accumulate in the millennia following the adoption of agriculture in some areas, but there is no strong

relationship between the years of agricultural dependency elapsed and the level of social complexity attained. Figure 5 provides a clear illustration of the dangers of focusing solely on traditions where social complexity has accumulated over time. If one did so in this instance, one might conclude that complexity is an inevitable result of agricultural dependence. Yet when all archaeological traditions are considered, it becomes clear that agricultural dependence is only one of many factors that have contributed to the accumulation of complexity. In other words, these data suggest that the accumulation of social complexity requires a continuous process of innovation, in both the physical and social realms.

The pattern discussed above shows that the accumulation of social complexity has been uneven, has not necessarily benefited group health, has not been strongly shaped by the environment, and is not an inevitable result of the agricultural revolution. Exploration of these data using multiple linear regressions also shows that there are no combinations of these variables that predict largest settlement populations in a statistically significant way. We therefore conclude that the accumulation of social complexity has not been driven by the passage of time, group-level biological benefits, environmental productivity, or the onset of agriculture. The determinants of this process need to be sought in other areas.

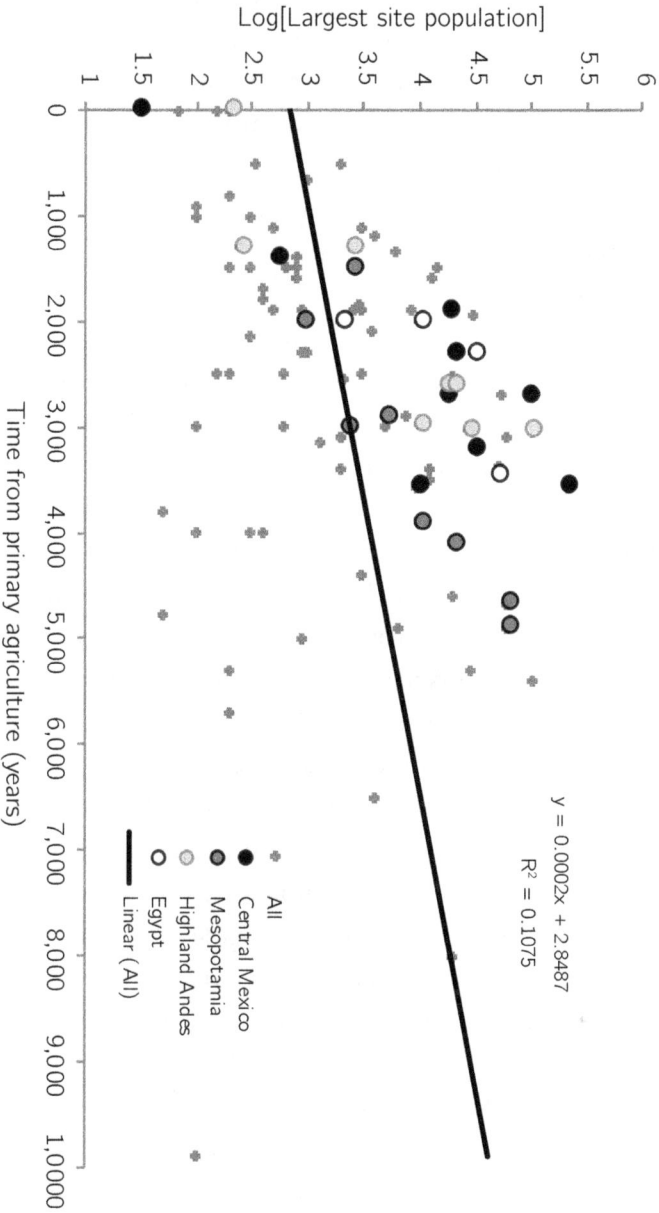

FIGURE 5 Time from primary agricultural dependence vs. largest settlement population. Traditions from areas of primary state formation are highlighted.

Economies of Scale

We now turn to several strong patterns that do emerge from these data. One is the notion of economies of scale. Several recent studies of contemporary urban systems have demonstrated that larger cities use resources more efficiently per capita, and produce more per capita, than smaller ones (Bettencourt 2013; Bettencourt, Lobo, Helbing et al. 2007; Bettencourt, Lobo, and Strumsky 2007; Bettencourt et al. 2010). Initial studies of this phenomenon in an archaeological context suggest that these advantages may be universal (Ortman et al. 2014; Ortman et al. 2015). Figure 6, which compares the population vs. settled area of largest settlements from various archaeological traditions, provides additional support for this view. Note that in this case both data series are plotted on logarithmic scales; the solid line represents a power function fitted to these data, and the dashed line represents the relationship that would be apparent if settled area increased linearly with population (i.e., with an exponent of 1 instead of the observed 0.86).

~200~

Figure 6 suggests that, on average, individuals require less area per person as settlements grow in population. This, in turn, suggests that resources are utilized more efficiently by larger aggregates, and that larger groups are likely to produce more per capita on average. This pattern has previously been observed across settlements within single systems, but here we show this pattern across settlements from different archaeological traditions with widely varying economies and political organizations, which suggest that economies of scale are intrinsic and universal.

The implications of general, scale-invariant economies of scale are profound. Many studies of social evolution have adopted a Marxist-inspired position in which complexity benefited a small group of ruling elites in a substantial way but made life worse for most of the population (Flannery and Marcus 2012; Yoffee 2005). The lack of relationship between social complexity and health

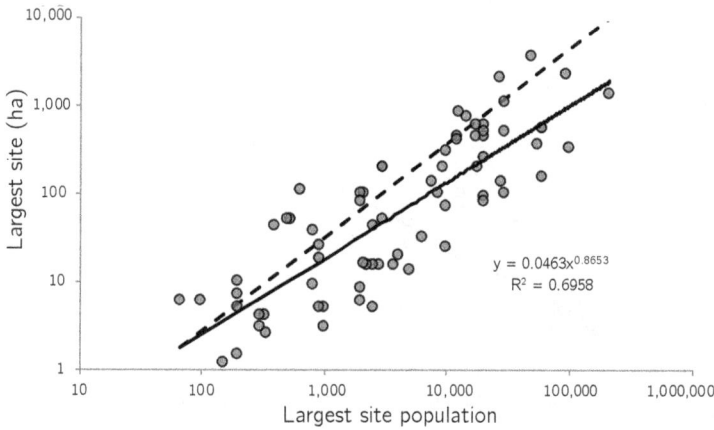

The chart shows Largest site (ha) on the y-axis (logarithmic scale from 1 to 10,000) versus Largest site population on the x-axis (logarithmic scale from 10 to 1,000,000).

$$y = 0.0463x^{0.8653}$$
$$R^2 = 0.6958$$

FIGURE 6 Population vs. settled area for the largest settlement in each archaeological tradition. The solid line represents the best-fit and the dashed line proportionate (linear) scaling.

discussed earlier reinforces this idea, or at least suggests that groups did not benefit biologically overall from increases in the scale of organization. Accordingly, such studies often conclude that the only way for complexity to emerge is for the ruling elites to build false consciousness among the commoner population through ideological means.

That elites in early civilizations promoted the virtues of large-scale political order is obvious, but what Figure 6 suggests is that these political ideologies were not pure fiction. If it is in fact the case that social networks exhibit open-ended economies of scale, then in any given environment larger-scale societies do in fact produce more per person, using less per person, than smaller-scale societies. The issue is not whether there are general benefits to scale, but how the benefits of scale are distributed through the social network. Many studies have shown that surplus production was not distributed evenly in early civilizations (Trigger 2003), and modern societies are not much better (Picketty and Saez 2014). But the potential for increases in scale to improve the average

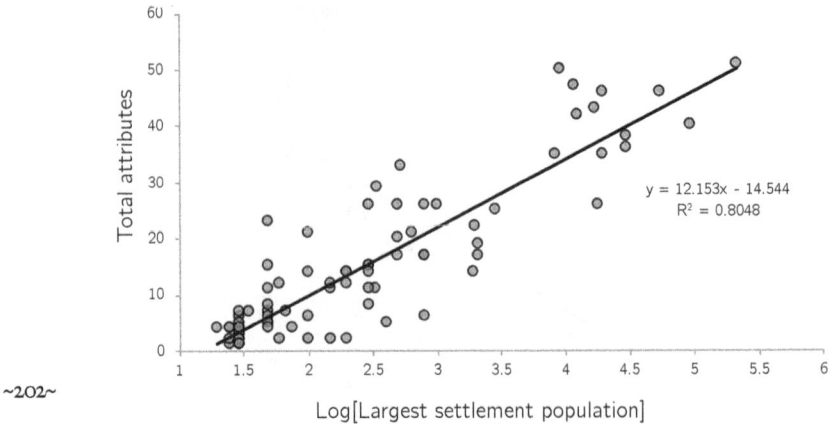

The chart shows a scatter plot with a fitted line:

$$y = 12.153x - 14.544$$
$$R^2 = 0.8048$$

Y-axis: Total attributes (0 to 60)
X-axis: Log[Largest settlement population] (1 to 6)

FIGURE 7 Relationship between scale (largest settlement population) and diversity of functions (total archaeological attributes) for a sample of North American and Mesoamerican traditions.

material conditions of life appears to have always been there, and it is not difficult to imagine that in some situations conditions did improve somewhat for much of the population, even if some benefited much more than others. Thus, we feel a more productive way to view the role of political discourse in the accumulation of social complexity is as arguments in favor of increasing the scale of social coordination and the benefits this can at least potentially bring to individuals within the group.

Functional Diversity

A second strong pattern we observe is that larger-scale societies support a greater diversity of functions. This is apparent in Figure 7, which compares the scale of social organization (the largest settlement population) with the total number of attributes from Table 1 that are associated with that archaeological tradition. The societies involved in this comparison all derive from North and Central

America. Although the list of attributes in Table 1 provides only a coarse representation of the full range of goods and services produced by a society, they nevertheless provide a consistent scale against which to compare societies quantitatively. The chart shows a strong positive relationship between the logarithm of the largest settlement population and diversity of functions, but the *rate* of increase of functional diversity decreases as settlements get larger. Recent studies of the diversity of professions in modern urban systems have identified a similar pattern of sublinear increase in occupational diversity with population across settlements within a single system. Importantly, this pattern is independent of the level of classification used to measure diversity (Bettencourt et al. 2014). This then appears to be an additional advantage of scale and an interesting property of cultural macroevolution: larger-scale societies generally support a wider range of goods and services, but the division of labor increases more slowly than the scale of organization.

Energy Capture

A third strong pattern we emphasize using a separate data source is the relationship between the scale of organization and the ability of a society to capture energy from the environment. Using documentary and archaeological data, Morris (2010) developed data series representing long-term changes in the rate of energy capture per capita for food, industry, commerce, and transportation in Western and Eastern civilizations. He also independently compiled estimates of the largest settlement populations in both civilizations, from 16,000 BP in the West and from 6,000 BP in the East. Although there are caveats to the data, they nevertheless represent a serious attempt to quantify two distinct dimensions of human social development over many millennia. Figure 8 presents these data for the East and West, taken from the online appendix to Morris's book (http://www.ianmorris.org/socdev.html).

Figure 8 shows that in both civilizations the scale of social organization increased much more rapidly and with much greater volatility over time than per capita energy capture did. However, it can be shown that the two data series are related in a deeper way. Settlement scaling theory suggests the average social connections of an individual, k, can be estimated from the size of the population aggregate, N, according to $k = k_0 N^{1/6}$, where k_0 is the average number of social connections per capita in the smallest settlement in the system (Bettencourt 2013, 2014). Each individual in a settlement also requires a minimum amount of energy per day, and as settlements grow individuals must receive an increasing fraction of their daily caloric requirements through their social connections, k. Some of the energy coming in to a settlement is lost with each interaction, and as a result energy dissipation scales with the same exponent as connectivity. Thus, the largest possible settlement that can be maintained at a given rate of energy input is given by $N \leq (E/w_0)^{6/7}$, where E is the total energy flowing into a settlement per unit time and w_0 is the baseline energy consumption rate (essentially the basal metabolic rate in early societies). This relation can be rearranged to yield $E \geq w_0 N^{7/6}$, which implies $e \geq w_0 N^{1/6}$, where e is the per capita energy capture rate. Thus, the largest settlement population values can be transformed via this relation into a minimum implied energy capture rate. These estimates of e derived from N, shown as dashed lines in Figure 8, closely track the energy capture rates estimated by Morris from the beginning of the data series up to the onset of the industrial revolution. This suggests that energy capture rates have been closely tied to the scale of social organization for most of human history. There is likely a feedback relationship between the two—larger social aggregates require greater rates of energy capture for their energy

(A) Western civilizations

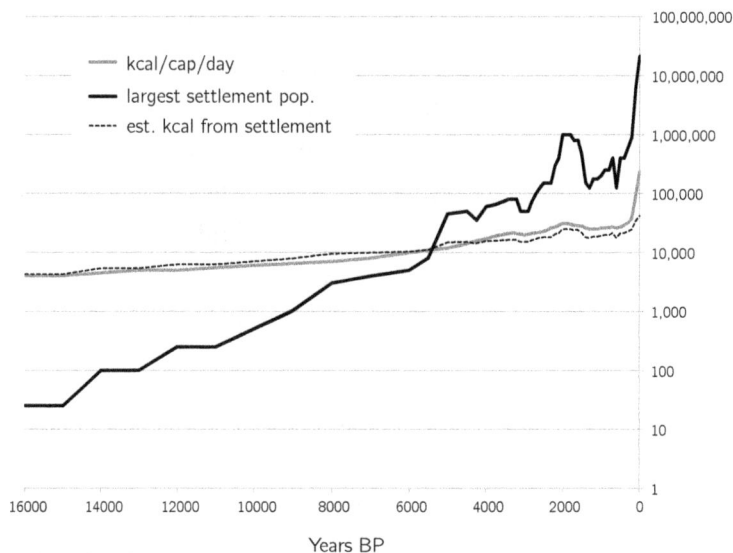

(B) Eastern civilizations

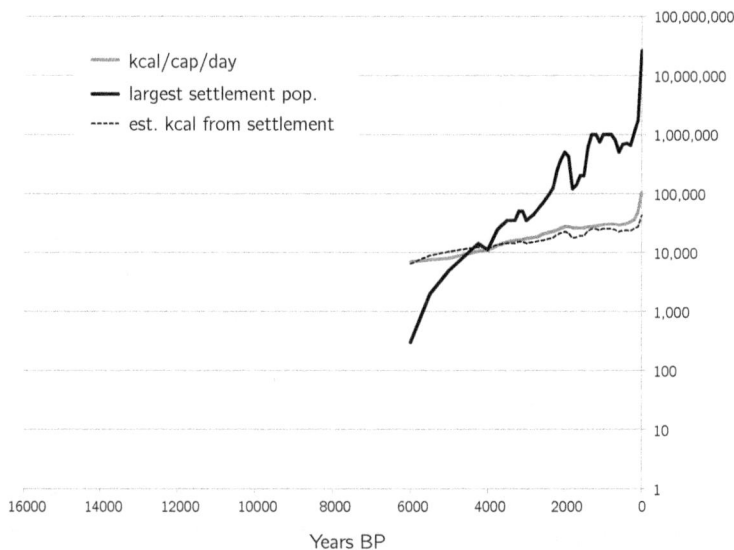

FIGURE 8 Energy capture and social organization in Western (A) and Eastern (B) civilizations, 16,000 BP to the present. In both charts, the dashed line represents $e_t = w_0 N_t^{1/6}$, where e_t is an estimate of energy capture rate per capita at time t, w_0 is the minimal energy requirement of an individual (2,500 kcal/day), and N_t is the population of the largest settlement at time t.

needs to be met given their concentration in space, and larger con-
centrations of people and interactions facilitate the expansion and
maintenance of the knowledge and technology for energy capture.
And in the same way, reductions in the scale of social organization
reduce the society's capacity to maintain knowledge, and societies
cannot maintain themselves at a given scale if the rate at which they
capture energy declines. Thus, it appears a third strong pattern in
cultural macroevolution is a systematic relationship between the
scale of social organization and energy capture.

~206~

A Cumulative Process

A fourth pattern that is apparent in the expanded ACE is that the
pace of change in human societies has been increasing over time
and is attributable in part to increases in the scale of organization.
Previous studies have shown that the characteristics of societies in
the ethnographic record can be arranged according to a Guttman
or implicational scale, such that societies at any given scale of organization typically possess nearly all the attributes of smaller-scale
societies (Gell-Mann 2011; Peregrine et al. 2004). This finding supports the hypothesis that cultural macroevolution is a cumulative
process. What this means is that, over time, societies tend to accumulate a stock of ideas, which can be combined in increasing numbers of ways in generating new ideas. This combinatorial process
should result in an exponentially increasing rate of cultural change,
and this increasing rate should be reflected in our data. We measure the rate of cultural change over time by examining patterns in
the duration of archaeological traditions. Each tradition is associated with a beginning and an ending date, and thus encompasses
a specific period of time. This duration represents the length of
time needed to accumulate enough social and cultural changes to
warrant the definition of a new tradition. The characteristics used
to distinguish two traditions can vary from major technological

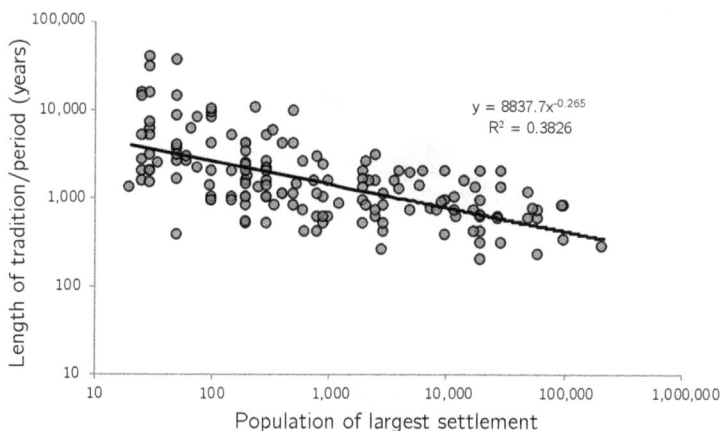

$$y = 8837.7x^{-0.265}$$
$$R^2 = 0.3826$$

FIGURE 9 Scale vs. duration of archaeological traditions.

innovations to the establishment of a new art style to the founding of a new political capital to societal collapse. But regardless of the details, the durations of archaeological traditions provide a coarse-grained measure of the overall rate of cultural change, in precise analogy to the way paleobiologists use patterns in the duration of genera in the fossil record to track changes in evolutionary rates over geologic time (Sepkoski 2012).

Figure 9 compares the duration of archaeological traditions against the population of their largest settlements, and Figure 10 compares the duration of traditions against their midpoints in time. These charts show that larger-scale societies have generally changed more rapidly than smaller-scale societies, but the correlation between duration and midpoints in time is even stronger. Furthermore, it appears that the average rate of increase in the rate of change has been fairly consistent from the Middle Stone Age (ca. 120,000 BP) onward. One could argue that this latter pattern is merely due to the fact that a wider range of more precisely

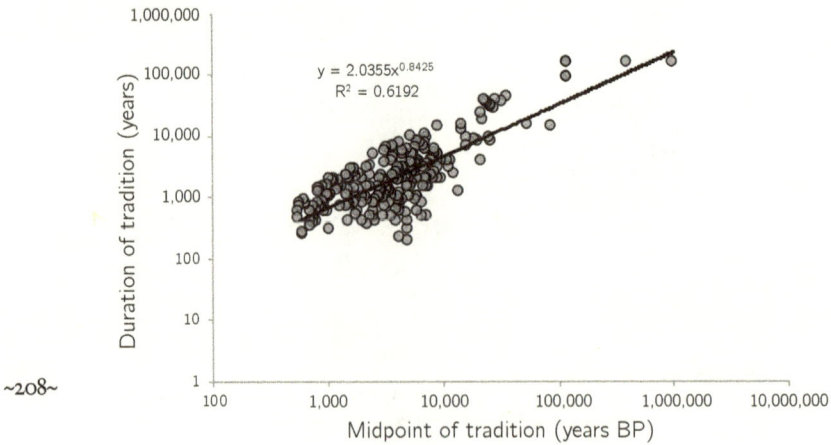

FIGURE 10 Midpoint vs. duration of archaeological traditions, Middle Stone Age to the present.

datable materials occurs on more recent sites. But this is precisely the point: the pace of change has accelerated over time in part because there have come to be more cultural ideas, more potential combinations of ideas, and thus a greater number of potentially preserved artifacts (Kline and Boyd 2010; Kuhn 2012; Powell et al. 2009). So these changes are not just easier to see, but also they reflect the accumulation of knowledge over time. Taken together, Figures 9 and 10 suggest the rate of cultural macroevolution has increased over time (also see Perreault [2012] for a comparison of cultural vs. biological evolutionary rates). This appears to be due in part to increases in scale, which we have already shown are associated with increasing productivity, connectivity, functional differentiation, and energy capture rates. But it is also likely to be due to the accumulation of ideas, and useful combinations of ideas, in humanity as a whole.

A final pattern we observe is that, although social evolution has a cumulative character, the scale of organization of human societies does not vary continuously. Figure 11 illustrates this in the form of a histogram of the logarithm of largest settlement populations across 157 archaeological traditions. Note that the distribution is multi-modal, with peaks corresponding to settlements of 25,500, 2,500, and more than 10,000 people. These peaks correspond closely to thresholds in human social organization observed in cross-cultural studies (Kosse 1990, 1992, 1994, 2001) and to traditional neo-evo-lutionary typologies of human societies (band, tribe, chiefdom, state) (Fried 1967; Service 1962). Although we do not favor the return to a typological approach, these data nevertheless suggest ~209~ that there are basins of attraction for various scales of human social

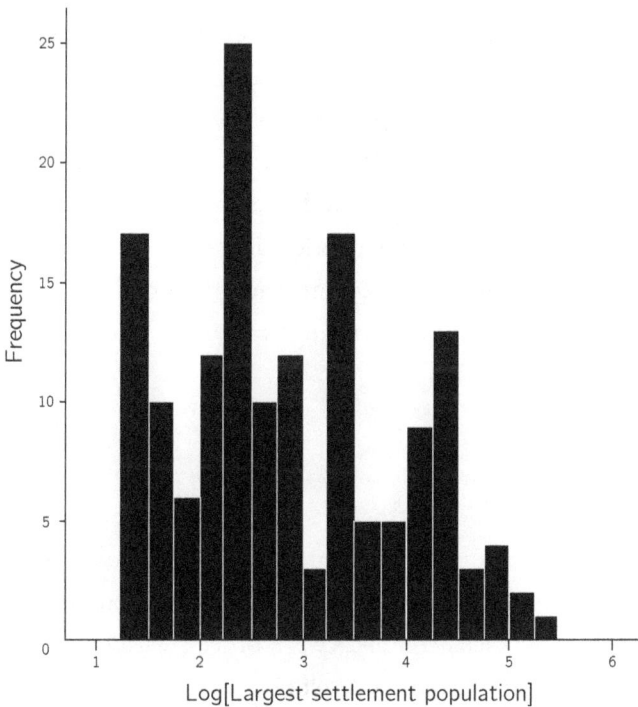

FIGURE 11 Histogram of largest settlement populations across archae-ological traditions.

organization and that these basins correspond in some sense to the neo-evolutionary types recognized many decades ago.

This multimodal pattern has been recognized for decades, but it is somewhat mysterious given the evidence for the consistent and continuous benefits of scale adduced in this chapter. Given these benefits, one might expect continuous feedback processes to have resulted in a continuous distribution of organizational scales. Yet this is not what has occurred, and the multimodal pattern in the scale of organization of human societies suggests there are additional constraints that make some scales of organization easier to attain and maintain than others. A good theory of cultural macroevolution needs to account for these thresholds, whether they derive from human information-processing limits (Kosse 1994), decision-making hierarchies (Johnson 1982), scalar limits in the effectiveness of different organizational concepts (clan, community, ethnic group, nation) (Kowalewski 2013), or some combination of these and other factors. Regardless of the specific explanations, the basic pattern shows that there are constraints upon the scale of human social organization that do not derive from the environment, economies of scale, the division and specialization of labor, energy capture rates, or the accumulation of knowledge, since all these factors show continuous change with scale or over time.

~210~

Summary

In this chapter we have summarized a database of information on archaeological traditions across the entire sweep of human prehistory to characterize quantitative patterns in human cultural macroevolution. The quantitative summaries presented here suggest the accumulation of social complexity is a process with the following dimensions: (1) an expanding envelope over time; (2) benefits that are not reducible to group-level biological benefits; (3) systematic economies of scale; (4) increases in functional diversity; (5)

increasing rates of social connectivity and energy capture; (6) cumulative growth of knowledge leading to increasing rates of culture change; and (7) multivariate basins of attraction for the scale of social organization.

Many of these characteristics are not new or surprising—as we have noted, anthropologists have been aware of relationships between the scale of social organization and social complexity for several decades. Correlates of scale that have been identified previously include the relationship between scale of organization and multivariate measures of complexity, and the multimodal distribution of organizational scales across societies. We have also identified correlates of scale that have not been noted previously, including systematic economies of scale in the use of resources and of productivity; a sublinear relationship between scale and functional diversity; a systematic relationship between organizational scale and energy capture; and a relationship between scale and the rate of cultural evolution. Thus, our study replicates earlier findings but also extends these findings using continuous measures of scale and time series data, which permit an analysis of change through time.

Readers may not be surprised to note the variety of correlates of organizational scale we have identified, as many studies of emergent complexity in individual societies have noted the appearance of new properties during this process. What we feel is more important, though, is the consistent rate at which various properties emerge across scales of organization. Many studies have appealed to specific processes—specialization, exchange, social stratification, religion, ideology, technological progress, and so forth—as causal factors in the emergence of social complexity in specific contexts. A critical question for future research is how to square the idea that specific processes are responsible for specific changes in individual societies with the strong patterns in cultural macroevolution identified in this study. We feel that a productive way to think about this issue is to recognize that there are a variety of specific social, cultural, and political ideas, institutions, and technologies in specific contexts that lead to similar functional results, and it is the net result of these details that lead to strong patterns in cultural macroevolution. So, for example, societies vary dramatically in the range of transportation technologies they possess, but all of these play the same functional role in a society by determining the energetic costs of moving people and goods across the landscape, and within and between settlements. Also, societies have developed a range of

social technologies that govern flows of goods and services—from gift-giving and barter to markets to money, and from finance and tithing to taxation and tribute—but all of these likewise play the same functional role in facilitating flows of energy and information through social networks. Finally, societies have devised a remarkable range of institutions that structure social networks—from kinship to communities to ethnic groups to nations to a whole range of other identities defined by biology, culture, religion, or occupation—but all of these play the same role in structuring human connectivity in space and time.

Our point here is that the strong patterns in cultural macroevolution suggest there should be a way of characterizing human societies as networks of matter, energy, and information that is context independent, such that the rich local variation identified and described through archaeological and ethnographic studies can be thought of as factors that influence the values of a few key parameters that determine the overall scale, structure, and functioning of the system. We find this to be a promising way forward, as it creates a role for studies that further investigate the strong regularities in cultural macroevolution and their functional parameters, and for studies that investigate how the ideas, institutions, and technologies of specific societies influence these basic functional parameters. In our view, integrating the traditional focus of archaeology on understanding the historically contingent fates of individual societies with a complex systems perspective on the regularities in cultural macroevolution is important and necessary if the archaeological record—the most extensive compendium of human experience in existence—is to play a larger role in the social sciences. Studies of individual societies provide a sense of what can happen and feed one's imagination regarding possible futures, whereas a macroevolutionary approach provides a sense of what tends to happen as the net result of diverse local details. Both are needed, but in recent decades the coarse-grained levels at which regularities in human affairs are most apparent have played a markedly secondary role. The results of this study suggest that a more balanced approach, which forges stronger links between the contours of cultural macroevolution and sequences of historically contingent change, is necessary for the continued maturation of archaeology as a social science. 🌿

~213~

REFERENCES CITED

Adams, Robert McC.
   1966   *The Evolution of Urban Society*. Aldine, Chicago.
Arthur, W. Brian
   2009   *The Nature of Technology: What It Is and How It Evolves*.
   Free Press, New York.
Bettencourt, Luis M. A.
   2013   The Origins of Scaling in Cities. *Science* 340:1438–1441.
   2014   Impact of Changing Technology on the Evolution
   of Complex Informational Networks. *Proceedings of the IEEE*
   102(12):1878–1891.
Bettencourt, Luis M. A., J. Lobo, D. Helbing, C. Kühnert, and G. B. West
   2007   Growth, Innovation, Scaling, and the Pace of Life of
   Cities. *Proceedings of the National Academy of Science of the U.S.A.*
   104:7301–7306.

Bettencourt, Luis M. A., J. Lobo, and D. Strumsky
2007   Invention in the City: Increasing Returns to Patenting as a Scaling Function of Metropolitan Size. *Research Policy* 36:107–120.

Bettencourt, Luis M. A., Jose Lobo, Deborah Strumsky, and Geoffrey B. West
2010   Urban Scaling and Its Deviations: Revealing the Structure of Wealth, Innovation and Crime across Cities. *PLoS ONE* 5(11):e13541.

Bettencourt, Luis M. A., Horacio Samaniego, and HyeJin Youn
2014   Professional Diversity and the Productivity of Cities. *Scientific Reports* 4:5393. DOI:5310.1038/srep05393.

Blanton, Richard E., and Lane Fargher
2008   *Collective Action in the Formation of Pre-Modern States.* Springer, New York.

Boquet-Appel, Jean-Pierre
2002   Paleoanthropological Traces of a Neolithic Demographic Transition. *Current Anthropology* 43(4):637–650.

Boquet-Appel, Jean-Pierre, and Ofer Bar-Yosef (editors)
2008   *The Neolithic Demographic Transition and Its Consequences.* Springer, Berlin.

Boquet-Appel, Jean-Pierre, and Stephen Naji
2006   Testing the Hypothesis of a Worldwide Neolithic Demographic Transition: Corroboration from American Cemeteries. *Current Anthropology* 47(2):341–365.

Brown, James H.
1995   *Macroecology.* University of Chicago Press, Chicago.

Carneiro, Robert L.
1962   Scale Analysis as an Instrument for the Study of Cultural Evolution. *Southwestern Journal of Anthropology* 18:149–169.
1967   On the Relationship Between Size of Population and Complexity of Social Organization. *Southwestern Journal of Anthropology* 23:234–243.
2000   The Transition from Quantity to Quality: A Neglected Causal Mechanism in Accounting for Social Evolution. *Proceedings of the National Academy of Science of the U.S.A.* 97(23):12926–12931.

Chick, Garry
1997   Cultural Complexity: The Concept and Its Measurement. *Cross-Cultural Research* 31(4):275–307.

Feinman, Gary
2011   Size, Complexity and Organizational Variation: A Comparative Approach. *Cross-Cultural Research* 45(1):37–59.

Feinman, Gary, and Jill Neitzel
1984   Too Many Types: An Overview of Sedentary Prestate
Societies in the Americas. In *Advances in Archaeological Method and
Theory, vol. 7*, edited by Michael B. Schiffer, pp. 39–102. Academic
Press, New York.

Flannery, Kent V., and Joyce Marcus
2012   *The Creation of Inequality: How Our Prehistoric Ancestors
Set the Stage for Monarchy, Slavery, and Empire*. Harvard University
Press, Cambridge, Massachusetts.

Fried, Morton H.
1967   *The Evolution of Political Society: An Essay in Political
Anthropology*. Random House, New York.

Gavrilets, Sergey, and Jonathan B. Losos
2009   Adaptive Radiation: Contrasting Theory with Data. *Science*
323:732–737.

Gavrilets, Sergey, and Aaron Vose
2005   Dynamic Patterns of Adaptive Radiation. *Proceedings of the
National Academy of Science of the U.S.A.* 102(50):18040–18045.

Gell-Mann, Murray
2011   Regularities in Human Affairs. *Cliodynamics* 2(1):52–70.

Gould, Stephen J.
1989   *Wonderful Life: The Burgess Shale and the Nature of
History*. W. W. Norton, New York.

Johnson, Allen W., and Timothy Earle
1987   *The Evolution of Human Societies: From Foraging Group to
Agrarian State*. Stanford University Press, Stanford, California.

Johnson, Gregory A.
1977   Aspects of Regional Analysis in Archaeology. *Annual
Review of Anthropology* 6:479–508.
1980   Rank-Size Convexity and System Integration: A View from
Archaeology. *Economic Geography* 56(3):234–247.
1982   Organizational Structure and Scalar Stress. In *Theory and
Explanation in Archaeology: The Southampton Conference*, edited by
Colin Renfrew, Michael J. Rowlands, and Barbara Abbot Segraves,
pp. 389–421. Academic Press, New York.
1987   The Changing Organization of Uruk Administration on
the Susiana Plain. In *The Archaeology of Western Iran: Settlement and
Society from Prehistory to the Islamic Conquest*, edited by Frank Hole,
pp. 107–140. Smithsonian Institution Press, Washington, DC.

~215~

Kline, Michelle A., and Robert Boyd
2010   Population Size Predicts Technological Complexity in Oceania. *Proceedings of the Royal Society B.* DOI:10.1098/rspb.2010.0452.

Kohler, Timothy A., and Kelsey M. Reese
2014   Long and Spatially Variable Neolithic Demographic Transition in the North American Southwest. *Proceedings of the National Academy of Science of the U.S.A.* 111(28):10101–10106.

Kosse, Krisztina
1990   Group Size and Societal Complexity: Thresholds in the Long-Term Memory. *Journal of Anthropological Archaeology* 9:275–303.
1992   Middle Range Societies from a Scalar Perspective.
1994   The Evolution of Large, Complex Groups. A Hypothesis. *Journal of Anthropological Archaeology* 13:35–50.
2001   Some Regularities in Human Group Formation and the Evolution of Societal Complexity. *Complexity* 6(1):60–64.

Kowalewski, Stephan A.
2013   The Work of Making Community. In *From Prehistoric Villages to Cities: Settlement Aggregation and Community Transformation,* edited by Jennifer Birch, pp. 201–218. Routledge, New York.

Kuhn, Steven L.
2012   Emergent Patterns of Creativity and Innovation in Early Technologies. *Developments in Quaternary Science* 16:69–87.

Morris, Ian
2010   *Why the West Rules—For Now.* Farrar, Straus and Giroux, New York.

Naroll, Raoul
1956   A Preliminary Index of Social Development. *American Anthropologist* 56:687–715.

Ortman, Scott G., Andrew H. F. Cabaniss, Jennie O. Sturm, and Luis M. A. Bettencourt
2014   The Pre-History of Urban Scaling. *PLoS ONE* 9(2):e87902. DOI:87910.81371/journal.pone.0087902.

Ortman, Scott G., Andrew Cabaniss, Jennie O. Sturm, and Luis M. A. Bettencourt
2015   Settlement Scaling and Increasing Returns in an Ancient Society. *Science Advances* 1e00066. DOI:10.1126/sciadv.00066.

Pauketat, Timothy R.
2007   *Chiefdoms and Other Archaeological Delusions.* AltaMira Press, Lanham, Maryland.

Peregrine, Peter N.
2003   Atlas of Cultural Evolution. *World Cultures* 14(1):2–88.

Peregrine, Peter N., Carol R. Ember, and Melvin Ember
2004   Universal Patterns in Cultural Evolution: An Empirical Analysis Using Guttman Scaling. *American Anthropologist* 106(1):145–149.

Perreault, Charles
2012   The Pace of Cultural Evolution. *PLoS ONE* 7(9):e45150.

Picketty, Thomas, and Emmanuel Saez
2014   Inequality in the Long Run. *Science* 344:838–843.

Powell, Adam, Stephen Shennan, and Mark G. Thomas
2009   Late Pleistocene Demography and the Appearance of Modern Human Behavior. *Science* 324:1298–1301.

Sallan, Lauren Cole, and Michael I. Coates
2010   End-Devonian Extinction and a Bottleneck in the Early Evolution of Modern Jawed Vertebrates. *Proceedings of the National Academy of Science of the U.S.A.* 107(22):10131–10135.

Sepkoski, David
2012   *Rereading the Fossil Record: The Growth of Paleobiology as an Evolutionary Discipline.* University of Chicago Press, Chicago.

Service, Elman
1962   *Primitive Social Organization.* Random House, New York.

Smith, Adam T.
2003   *The Political Landscape: Constellations of Authority in Early Complex Polities.* University of California Press, Berkeley.

Steckel, Richard H., and Jerome C. Rose (editors)
2002   *The Backbone of History: Health and Nutrition in the Western Hemisphere.* Cambridge University Press, Cambridge.

Trigger, Bruce G.
2003   *Understanding Early Civilizations.* Cambridge University Press, Cambridge.

Wood, J. W., G. R. Milner, H. C. Harpending, and K. M. Weiss
1992   The Osteological Paradox. *Current Anthropology* 33:343–370.

Yoffee, Norman
2005   *Myths of the Archaic State: Evolution of the Earliest Cities, States and Civilizations.* Cambridge University Press, Cambridge.

PART III
Syntheses

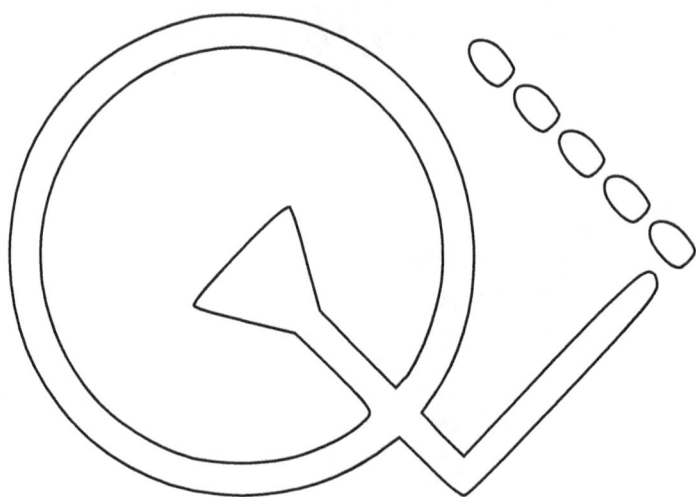

⌂

CULTURAL GENOTYPES AND
SOCIAL COMPLEXITY

Scott G. Ortman, University of Colorado Boulder and Santa Fe Institute

Human societies are complex networks of people, energy, and information, but our understanding of how humans actually represent information and utilize it in structuring behavior remains poorly developed. In this chapter I build a model of knowledge representation that is grounded in experimental research in cognitive and social psychology and which exhibits several key properties of genotypes noted in post-neo-Darwinian studies of evolutionary innovation. This framework also leads to several insights concerning the evolution of social complexity and a potential research program for studying it in evolutionary terms.

Specifically, I address a curious situation with respect to the relationship between evolutionary theory and research on social complexity. Over the past several decades anthropologists have explored a variety of ways in which evolutionary theory, drawn largely from biology, might help to characterize social change through time (Boyd and Richerson 1985; Cavalli-Sforza and Feldman 1981; Durham 1991; Laland and Brown 2002; Shennan 2002). For example, a recent summary identified five distinct schools of evolutionary thought in anthropology (Laland and Brown 2002):

- *Sociobiology* investigates the biological dimension of prosocial behavior.
- *Human behavioral ecology* investigates how the environment influences behavioral predispositions.
- *Evolutionary psychology* investigates the evolution of psychological predispositions.

- *Cultural evolution* considers cultural inheritance through analogy with the gene.
- *Gene-culture coevolution* considers how genes and culture influence each other across generations.

Although these approaches have been productive in addressing a number of questions, they have not had much impact on research concerning the origins of complex societies. Recent studies have utilized comparative approaches (M. E. Smith 2012; Trigger 2003), collective action theory (Blanton and Fargher 2008), various strands of Marxist theory (Feinman and Marcus 1998; Pauketat 2007; A. Smith 2003; Yoffee 2005), concepts of agency (Flannery and Marcus 2012; M. L. Smith 2003), and even socioeconomic development theory (Morris 2010), but none have been grounded in the various strands of evolutionary thought listed above.

Why is this? There are probably many reasons, but here I would like to emphasize three factors related to the grounding of existing approaches in the neo-Darwinian synthesis of natural selection and population genetics, which was forged in the 1930s through 1950s. First, cultural evolutionary models inspired by the neo-Darwinian synthesis assert a relatively direct relationship between human behavior and the cognitive phenomena that govern it. In other words, these models lack a realistic model of the way information is represented and stored in individual minds and utilized in shaping individual behavior. As a result, evolutionary models have had some success in characterizing the evolution of technology (Henrich 2001, 2004) but have not been especially relevant for understanding the role of social, cultural, political, and religious processes in the emergence of complex societies. Second, neo-Darwinian models tend to abstract specific individual practices from the totality of sociocultural processes, leading to an atomistic view of human society that is difficult to square with our more interconnected experience of social life. As a result, these models have not proven very useful for understanding the emergent properties of complex societies. Third, neo-Darwinian models have been shown to have limitations for understanding the evolution of complexity in biology itself (Erwin and Valentine 2013; Krakauer 2011; Maynard Smith and Szathmary 1999; Simpson 2011, 2012). So if these models cannot explain why there are primates as well as paramecia, they are not likely to be useful for explaining why there are cities as well as hunter-gatherers. For these reasons, it seems necessary to move beyond neo-Darwinian approaches if evolutionary thought is to play a larger role in the study of social complexity.

Fortunately, evolutionary biology is already moving in this ~223~ direction, and the view of evolution that is emerging provides a productive basis for reformulating the cultural evolutionary process in a way that is much more appropriate for the study of social complexity. In this chapter I seek to move the discussion in this direction by integrating three streams of research. First, I incorporate the latest insights from evolutionary biology and cognitive science in a new model of *cultural genotypes*. Second, I combine ideas from complex systems, economics, and social psychology to define how shared ideas enable the growth of social networks. Third, I outline methods through which one can track the evolution of cultural genotypes and quantify their large-scale behavioral effects through combined archaeological and cognitive-historical linguistic research. My basic thesis is that the resistance of many anthropologists to evolutionary approaches to culture change ultimately derives from limitations of neo-Darwinian models of evolution. I suggest that reformulating the cultural evolutionary process in light of current understandings in biology and related fields leads to models that are much more compatible with the empirical record of social evolution, and may lead to a new and more productive relationship between evolutionary theory and anthropology.

TABLE 1 Examples of genotypes and phenotypes from biological systems (after Wagner 2011:Figure 5.1).

System	Genotype	Phenotype
DNA	Sequence of base pairs in the DNA molecule	Sequence of coded amino acids
Proteins	Sequence of amino acids that comprise the molecule	Shape (and function) of the resulting proteins
Metabolic networks	List of all enzymes an organism can produce	List of nutrients the organism can metabolize
Gene regulatry networks	Matrix of gene interactions	The expression state at equilibrium

Insights from Biology

Three basic insights from recent evolutionary biology open up the space for a post-neo-Darwinian evolutionary anthropology. The first is that information is stored at all levels of biological organization. The neo-Darwinian synthesis held that all the information involved in making a functioning organism was encoded in the sequence of DNA base pairs. This view is enshrined in the concept of the *gene* as the particulate unit of inheritance in population genetics and the *meme* concept in anthropology (Dawkins 1986). However, recent studies in evolutionary development and ecology demonstrate that information about the environment is also incorporated into, and interacts with, the proteins produced by DNA transcription, gene regulatory networks, metabolic reaction networks, and even ecological relationships among organisms and species (Carroll 2006; Erwin 2012; Erwin and Davidson 2009; Maynard Smith and Szathmary 1999; Odling-Smee et al. 2003). Such studies are leading to a more expansive view of *genotypes as logical structures that represent and store information.* Although the base-pair sequence of the genome is what is directly inherited in most biological lineages, most of the information contained in this sequence only becomes explicit through the biological structures— from proteins to gene regulatory networks to metabolic networks to body systems—that emerge during development. As a result, natural selection acts on the information incorporated into every level of the phenotype. Table 1 presents examples of genotypes and phenotypes from a recent treatment of this issue (A. Wagner 2011). The physical substrates used to represent information differ in each system, but their phenotypic effects are similar in each case.

The second insight is that there is a many-to-one relationship between genotypes and phenotypes. As examples: several different base-pair sequences code for the same amino acid; several different sequences of amino acid produce proteins with the same shape and ~225~ function; several different combinations of enzymes can be used to metabolize nutrients in a metabolic network; and many different patterns of gene regulation can produce the same overall expression in gene regulatory networks (Erwin and Davidson 2009; Ferrada 2014). This realization has led to the concept of a *genotype-phenotype map*, which defines all the logically possible relationships between genotypes and phenotypes. Such maps can be visualized as n-dimensional grids where each node represents the coordinates of a given genotype in sequence space, and there are regions of genotype space that correspond to a given phenotype (Figure 1). The fact that many genotypes produce the same phenotype allows species to explore genotype space while preserving viable phenotypes, thus increasing the chances that new viable phenotypes will be discovered.

In this framework, novelty emerges from three processes: (1) *expansion of the genotype space* through increases in the size or scope of the representation (e.g., the number of DNA base pairs, the number of metabolic reactions, etc.); (2) the *discovery of new combinations* of genotypic elements that lead to a viable phenotype; and (3) the *emergence of new levels* of genotype. The latter may involve the appearance of traits that are beneficial in the context of groups but harmful to individuals in isolation (Libby and

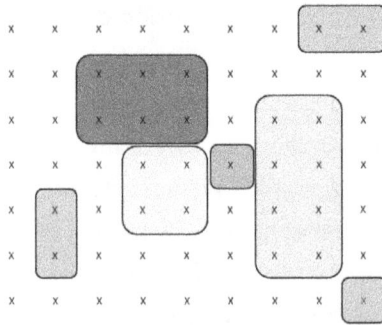

FIGURE 1 Visualization of a genotype-phenotype map. Each node corresponds to a specific genotype (e.g., an amino acid sequence) and adjacent nodes differ by a single element (one amino acid substitution). The shaded regions correspond to viable phenotypes. Biological lineages drift within each shaded region without losing viability, thus increasing the chances that a new viable phenotype will be discovered.

Ratcliff 2014) or increases in the reproductive division of labor among aggregates of lower-level units (Simpson 2012) (e.g., the nucleus of a eukaryotic cell, germ cells in multicelled organisms, queens in social insects, etc.).

The third insight is that information is hierarchically organized (Simpson 2011). The clearest example of this is gene regulatory networks that control gene expression during development (Erwin and Davidson 2009). A key element of such networks are highly conserved genes that control essential developmental processes, metabolic processes, and biological structures (Carroll 2006; Shubin 2009). These patterns demonstrate that levels of genotype that emerged earlier in the history of life are far less evolvable than those that emerged later, and as a result there is strong path dependency in evolution—like computer programs, genotypes tend to build on themselves instead of starting over (Erwin and Valentine 2013). The neo-Darwinian synthesis proposes that the only significant constraints on evolution are imposed by the environment, but recent studies demonstrate that genotypes also impose constraints, and many of these are legacies of evolutionary paths taken long ago (Gould 1989; G. Wagner 2014).

These three insights are driving a revolution in evolutionary biology, but they have yet to have an impact on models of cultural evolution. It should be clear from this summary that the primary impediment to the incorporation of these ideas is a good model of cultural genotypes—the physical means by which humans represent information in the brain and utilize it in shaping behavior. Traditional anthropological understandings suggest that such a model is within reach, given the new thinking in biology. Human ~227~ culture, as a system that represents, stores, and shares information derived from experience, is readily accommodated to the more flexible view of genotypes as logical information structures. The concept of the genotype-phenotype map also reflects a phenomenon first noted by anthropologists decades ago: namely, that many different cultural (emic) representations can lead to the same functional (etic) effects. New Guinea highlanders stage pig feasts in order to repay debts to their ancestors, but the net result of these ceremonies is improvement in the health and sustainability of the community (Rappaport 1968). In the same way, Tewa Pueblo ceremonies are intended to assist the growth of crops, but the net result of the associated feasting is to redistribute surpluses at times of year that people need it most (Ford 1972). Finally, recent work suggests that the same processes noted in studies of evolutionary innovation in biology—including expansion of the information space, the discovery of new combinations of existing elements, and the emergence of hierarchical relations—are also involved in technological progress (Arthur 2009; McNerney et al. 2011; Nagy et al. 2012; Youn et al. 2015). These correspondences suggest that it may now be possible to define cultural genotypes in a way that brings the realm of information into cultural evolutionary analysis.

Defining the Cultural Genotype

To propose that human culture is just another form of genotype is to say that, fundamentally, culture is a system for representing, storing, and retrieving information accumulated from past experience. But what does cultural information consist of? What are its units in physical terms? Cognitive neuroscience suggests that the fundamental units of human thought are *analogies* whose

~228~

What does cultural information consist of? What are its units in physical terms?

physical manifestations are neural activation patterns in individual brains (Feldman 2006). Indeed, such research advances the idea that analogies provide the "fuel and fire," or the raw material and reactions, of thinking (Hofstadter and Sander 2013). Here, I propose that the most highly developed area of research on analogy, known as conceptual metaphor theory (CMT), provides a good starting point for a model of cultural genotypes.

In contemporary cognitive science, *conceptual metaphor* is a label for the highly developed human ability to model relatively complex or abstract phenomena using the structure of simpler or more concrete phenomena. It also refers to the cognitive models resulting from this process that humans use in thinking, reasoning, and speaking. Conceptual metaphors arise in our brains from correspondences in the image-schematic structure of distinct experiential domains. Based on the principle that "neurons that fire together wire together," these image-schematic correspondences lead to the mapping of entities, properties, and relationships from the source domain onto the more abstract target domain (Feldman 2006; Kövecses 2002; Lakoff and Johnson 1980). This makes it possible for people to conceptualize and reason about the target domain in terms of the structure provided by the source domain.

For example, most readers know what I am doing when I "lay the foundations of my argument" because English speakers have internalized the metaphor ARGUMENTS ARE BUILDINGS, and no building can stand without a solid "foundation." If one thinks for a moment about the process of "constructing" an argument, one will see that architectural imagery and reasoning is central. When "building" an argument, one needs to identify its "parts" and the order in which they should be "put together." One also needs to ensure that the conclusions "rest squarely" on a "secure foundation" and that the argument "holds together," can "withstand" criticism, and will not "crumble under its own weight." It is possible to talk about intellectual arguments without using metaphorical language, but it usually takes more specialized vocabulary and more complex grammatical constructions to convey concepts that are much more directly expressed using the concrete terminology of buildings and other tangible objects. And even if one succeeds in avoiding metaphorical language, it is very difficult to think about intellectual arguments in the absence of more general metaphors like MENTAL PROCESSES ARE PHYSICAL PROCESSES, of which ARGUMENTS ARE BUILDINGS is an example. The key point, then, is that conceptual metaphors are not just poetic tropes used to communicate nonmetaphorical concepts. Rather, these image-schematic mappings are the concepts themselves.

Studies of natural language use demonstrate that people use hundreds of conventional (shared or cultural) metaphors unconsciously and automatically in everyday thinking and communicating. These widely shared metaphors are part of the cognitive unconscious of a social group and are fundamental to its worldview, language, and behavior. Research in CMT has shown that conceptual metaphors play important roles in language, literature, philosophy, psychology, mathematics, material culture, religion, and politics (M. Johnson 1987; Lakoff 1987, 1993; Lakoff and Johnson 1980, 1999; Lakoff and Kövecses 1987; Lakoff and Núñez 2000; Lakoff and Turner 1989; Ortman 2000, 2012; Palmer 1996; Shore 1996; Slingerland 2004; Sweetser 1990; Wiseman 2014). It has also been successful in promoting metaphor from the status of literary trope to a fundamental mechanism of abstract thought and a subject of significant experimental research. For example, recent experimental work demonstrates that temporal reasoning is influenced by spatial schemas (Boroditsky 2000), opinions about social issues are influenced by priming with body-state imagery (Landau et al. 2009), people's desire for cleaning products is increased by exposure to stories of immoral behavior (Zhong and Liljenquist 2006), exposure to stories of moral violations such as cheating on an exam arouse feelings of (moral) disgust and reduce oral consumption (Chan et al. 2014), exposure to ideas about supernatural beings induces people to behave as though they were being watched by actual people (Sharif and Norenzayan 2007), and the physical act of walking toward a picture of an acquaintance reduces the subjective feeling of social distance from that person (Travers 2015). This range of research demonstrates that CMT truly represents a fundamental advance in understanding of human thought, language, culture, and behavior. As Steven Pinker (2007:253–276) concludes:

> Metaphor really is a key to explaining thought and language. The human mind comes equipped with an ability to penetrate the cladding of sensory experience and discern the abstract construction underneath—not always on demand, and not infallibly, but often enough and insightfully enough to shape the human condition. Our powers of analogy allow us to apply ancient neural structures to newfound subject matter, to discover hidden laws and systems in nature, and not least, to amplify the expressive power of language itself. . . . conceptual metaphors

are not just literary garnishes but aids to reason—they are "metaphors we live by." And metaphors can power sophisticated inferences, not just obvious ones.

There are several reasons why CMT also provides a strong foundation for a model of cultural genotypes. First, it provides a model of the neural encoding system, consisting of image-schematic correspondences and mappings, which humans use to represent and store information and transmit it to others. There is a sense in which conceptual metaphors are similar to memes; but they differ in that the latter has typically been modeled as a particulate unit of ~231~ inheritance analogous to the gene, whereas the former is a network of image-schematic information with significant internal structure defined by the mapping between the source and target domains. Conceptual metaphors allow one to see inside the black box of human psychology because, although metaphors exist only as networks of neural activation, the imagery they connect ultimately derives from the world of perceptual and bodily experience. Thus, cultural genotypes are built from the image-schematic structure of human experience in ways researchers can perceive and analyze, even though it is currently difficult to directly observe their physical instantiations inside individual brains.

Second, systems of conceptual metaphors vary between human groups and have changed over time. Some metaphors, such as ANGER IS HEAT ("he blew up at me") and SOCIAL DISTANCE IS PHYSICAL DISTANCE ("we are closely related") are probably universal (or near universal) because they derive directly from innate bodily and emotional experience (see Kövecses 2002:163–182; Wiseman 2014). However, other metaphors are more culturally specific due to differences in history, environment, knowledge, and technology (think, for example, of computer viruses). Indeed, the source domains of all metaphors ultimately derive from human experience, and this implies that the realm of potential source domains for new metaphors has expanded along with human technology and social organization over time. In short, systems of metaphors, and thus cultural genotypes, can and do evolve.

Third, conceptual metaphors are readily combined and blended at different levels of abstraction, thus creating hierarchical and networked logical structures that lead to new and more complex ideas, technologies, and behaviors (Fauconnier 1997; Fauconnier and Turner 1994, 2002; Ortman 2000). Research has identified a number of rules that govern how this recombination process works, and these empirical generalizations provide an outline of how novelties in cultural genotypes originate (Lakoff 1993; Ortman 2000). In addition, there is a many-to-many relationship between conceptual structures and behavior, in the sense that the same objective behavior can emerge from a variety of different conceptualizations, and a single metaphor can encourage a variety of different concrete behaviors depending on which specific correspondences and mappings are emphasized.

CMT is also compatible with traditional views of human rationality because it provides a model of the concepts to which reason is applied. Indeed, it is important to recognize that reason is applied not with respect to the world itself but to representations of the world; and when these representations deal with complex social, ecological, economic, political, or technological phenomena, they consist primarily of conceptual metaphors. It is therefore best to think of metaphors as tools that enable the application of fast and intuitive thought, and slow and rational thought, to abstract domains (Haidt 2012; Kahneman 2011).

Finally, and perhaps most important, conceptual metaphors are the basic units from which religious and moral concepts are built (Haidt 2012; Norenzayan 2013; Slingerland 2008), and thus an understanding of how metaphors influence social behavior is key to understanding the cooperation among anonymous strangers that is so characteristic of large-scale human societies. The anthropological literature suggests that the metaphors that support large-scale social coordination often utilize experiential correlates of social distance, and especially biological relatedness, as a source domain. For example, Marshall Sahlins (1972:123) notes that fictive kinship—essentially using experience of family life as a source domain for conceptualizing a wider circle of people as biological relatives—plays an essential role in economic intensification beyond the domestic mode of production. In addition, people in many societies conceptualize forces of nature as parents who monitor social behavior in the "family" (Norenzayan 2013; Trigger 2003). In other societies, people utilize the experience of generalized reciprocity within a household to conceptualize the social organization of a settlement, and this is reflected in public buildings that take the form of scaled-up domestic residences (Hodder 1990; Ortman 2011a; Ur 2014). Finally, people often use experiential correlates of physical closeness to promote social coordination through public ritual (Durkheim 1968; Hegmon 1989; Rappaport 1979). Indeed, ritual relationships have been shown to influence interband interaction rates even in foraging societies (Hill et al. 2014). It therefore appears that the invention and spread of conceptual metaphors that reduce the psychological social distance between people who are not experientially close is an essential component of social evolution.

~233~

From Cultural Genotypes to Social Networks

Developments in evolutionary biology reviewed above create the space for a model of the cultural genotype, and research in cognitive science suggests that CMT provides a good starting point. The third essential ingredient of a post-neo-Darwinian cultural evolutionary framework is a useful model of a human society. For this third component, I draw upon the view of human societies as social reactors developed in urban scaling theory. Many studies over the last few decades have demonstrated that contemporary cities exhibit statistical regularities that describe how their properties— from the connectivity of social networks to socioeconomic outputs, and from the division of labor to land area and extent of infrastructure—vary systematically with population size (Angel et al. 2011;

These results are critical because they suggest that there are significant and intrinsic benefits of scale for human groups, and therefore many of the properties of complex societies identified in past archaeological research are simply nonlinear effects of scale in human social networks.

Batty 2008; Bettencourt 2013, 2014; Bettencourt, Lobo, Helbing et al. 2007; Bettencourt, Lobo, and Strumsky 2007; Bettencourt et al. 2014; Glaeser and Sacerdote 1999; Nordbeck 1971; Samaniego and Moses 2009; Schläpfer et al. 2014; Sveikauskas 1975). Generally, socioeconomic quantities, Y (such as wages, patents, or violent crime), increase, on average, faster than population and are well described by scale invariant relations, $Y = Y_0 N^\beta$, with $\beta = 1 + \delta \sim 7/6 > 1$, with $\delta \sim 1/6$. Conversely, measures of the physical extent of urban infrastructure (roads, pipes, settled area, etc.) increase only sublinearly with N, $\beta = 1 - \delta \sim 5/6 < 1$. Urban scaling theory proposes that these properties emerge from a few general properties of human social networks embedded in space (Bettencourt 2013), such that larger cities are environments where a larger number of social interactions per unit time can be supported and sustained. This generic dynamics is in turn the basis for expansions in the division and coordination of labor, the specialization of knowledge, and the development of (hierarchical) political and civic institutions.

An exciting outgrowth of this research is evidence that urban scaling patterns have characterized nonurban settlement systems throughout history and may well derive from the same fundamental processes (Ortman et al. 2014; Ortman et al. 2015; Ortman and Coffey 2017). Anthropologists have long been aware of the close relationship between the populations of human societies and their social, political, and economic complexity (Carneiro 1962, 1967, 2000; Feinman 2011; Gell-Mann 2011; Henrich 2004; Kline and Boyd 2010; Peregrine et al. 2004; Powell et al. 2009). Settlement scaling theory continues this line of research in proposing that these properties ultimately emerge from increased rates of social interaction, increasing flows of goods and services, and increased information storage capacity

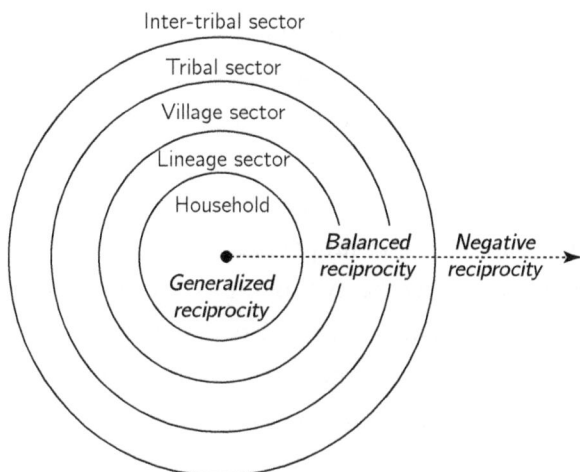

FIGURE 2 Relationship between social (in this case, kinship) distance and forms of exchange in small-scale societies (from Sahlins 1972:199). The exchange continuum is also a moral continuum concerning appropriate treatment of individuals at various perceived social distances.

of the social networks that develop when people live and work in close proximity.

These results are critical because they suggest that there are significant and intrinsic benefits of scale for human groups, and therefore many of the properties of complex societies identified in past archaeological research are simply nonlinear effects of scale in human social networks. But here is a critical point: despite these intrinsic benefits, only some societies have figured out how to take advantage of them. This is apparent from analyses of the expanded *Atlas of Cultural Evolution,* which show that small-scale archaeological traditions have continued to form until recent times, and that larger-scale societies are not a simple product of more productive environments (Ortman et al., chapter 7 in this volume). Ethnographic studies also show that social coordination problems (as opposed to resource limitations) are a primary reason small-scale societies stay small (Walker and Hill 2014). So even if settlement scaling theory helps to explain why the proportion of humanity living in larger-scale societies has grown over time, a good theory of social evolution must also explain why so many small-scale societies have continued to form, and what it is exactly that enables certain societies to take advantage of scale.

In light of the preceding discussion, I suggest that the most fundamental constraint on the emergence of social complexity is cultural genotypes that support the growth of densely interacting social networks. In other words, innate human instincts, by themselves, are not sufficient for a society to support robust networks of interacting anonymous strangers. Humans are inherently prosocial toward close kin, in the sense that people are generally willing to help close kin even at a cost to themselves. Researchers have proposed a variety of evolutionary processes, including kin selection, reciprocal altruism, and group selection, that appear adequate for explaining the emergence of these instincts in humans and other species (Bowles and Gintis 2011; Richerson and Boyd 2005; Singer 1981). However, these prosocial instincts are not automatically extended to socially distant persons. For example, in small-scale societies the dominant mode of exchange is correlated with kinship distance: generalized reciprocity (altruism) prevails among close kin, balanced reciprocity among more distant kin, and negative reciprocity among nonkin (Sahlins 1972:198–202). There is also an important correlation between kinship distance, forms of reciprocity, and moral norms, as it is generally viewed as morally wrong to take advantage of close kin, whereas it is morally appropriate to seek advantage from strangers (Figure 2).   ~237~

In contemporary US society, laws derived from social norms are intended to be applied equally to all citizens by political institutions that are disinterested in the specifics of the personae involved (Acemoglu and Robinson 2012; North et al. 2009). These universal and impersonal norms make social interactions among anonymous strangers much more effective and predictable, and in turn provide essential support for our large and complex economy. For people accustomed to life in such a society, it is difficult to imagine living in a world where it is viewed as appropriate to take advantage of strangers. And yet the ethnographic record shows that in most small-scale societies this is typical. Under these conditions large social networks, and everything that flows from them, cannot exist. So if the benefits of scale have always been there but humans are by nature cooperative only toward people with whom they are experientially close, the most fundamental determinant of social evolution must be shared ideas that support the extension of these innate cooperative instincts to experientially distant persons.

Steven Pinker (2011) reached a similar conclusion in his attempts to explain a related pattern: the gradual decline in violence over the course of human history. Pinker notes that although technology, resources, wealth, and religion have all influenced rates

THE EMERGENCE OF PREMODERN STATES

of within- and between-group violence in specific contexts, none of these factors are consistently associated with declines. He then identifies a variety of factors that are so associated and suggests that all of them work by changing the calculus of prisoner's dilemma games in favor of mutual cooperation, which implies increased trust. These factors include the growth of impersonal and institutionalized government; increasing economic integration; devaluation of male honor, bravery, and valor; expansion of the circle of sympathy; and the growth of abstract and universalizing reason (Pinker 2011:678–692). The first two factors have also been emphasized in the new institutional economics, but researchers in this field note that the political and economic institutions that create incentives for individual behavior ultimately derive from shared ideas, norms, and values (Acemoglu and Robinson 2012; North et al. 2009). The other three factors—devaluation of male aggressiveness, expanding empathy, and abstract reason—are also statements about ideas, norms, and values. So it appears that, in the end, the most important drivers of what Pinker calls the "arrow of history" are shared ideas, norms, and values concerning social relations.

The Fundamental Driver of Social Complexity

I am now in a position to state a hypothesis concerning the fundamental process behind the evolution of social complexity. Because shared ideas become increasingly abstract as they encompass larger and more impersonal groups, the fundamental driver of social evolution must be the invention and spread of conceptual metaphors concerning social relations. Furthermore, because humans are innately prosocial toward close kin, the metaphors most responsible for social evolution will most often utilize bodily experience of social closeness as the source domain for conceptualizing social relations with an expanded circle of people. These mappings interact with human moral instincts in defining social norms and values, and thus provide the foundations of political and economic institutions that expand the size of social networks.

In framing the situation this way, I do not intend to overlook the role of selection on phenotypes, as both natural selection and human choice are essential elements of social evolution, and there is a large volume of excellent work on both natural selection and choice in the context of humans and other social species (Bowles and Gintis 2011; Boyd and Richerson 1985; Damasio 1994; Frank 1998; Kahneman 2011). What I do seek to add to this dis- ~239~ cussion is a deeper understanding of the mechanisms by which novel cultural genotypes arise and the ways in which these novelties influence individual and interpersonal behavior, and thus social coordination in all its forms, from warfare to economics.

The somatic marker hypothesis, as developed by Antonio Damasio (1994), provides a good starting point for understanding how conceptual metaphors encourage social coordination. Damasio explains that humans and other animals possess a variety of automatic body-state responses to stimuli that have evolved for the purpose of maintaining the organism in homeostasis. These are known as primary emotions. When one of these responses is triggered, humans and other animals create mental images of the body-state response. These are secondary emotions. Bodily experience, then, consists of associations between external stimuli and body-state responses that are stored in long-term memory. These constellations of imagery can be recalled directly to facilitate reasoning about potential consequences of future courses of action, with the decision-making process involving evaluation of the body-state implications of similar past experiences. Here is the critical point: these bodily experiences, including secondary emotions, also form the source domains of conceptual metaphors, and when this happens the emotional correlates of the source domain are recruited to assist in reasoning about the more abstract target domain (Kövecses 2002). So, for example, when a soldier thinks of his platoon-mates as "brothers," he maps the positive secondary emotions of childhood, shared experience, teamwork, and mutual care onto his relations with his platoon-mates, thus encouraging him to treat them as he would actual siblings. I believe this is the fundamental mechanism behind social evolution.

Another way of stating this hypothesis is to follow Nowak (2006) in suggesting that the key to social evolution is an adaptation of Hamilton's rule. In biology, Hamilton's rule states that cooperative phenotypes will increase in a population when $rB - C > 0$, where r is the coefficient of genetic relatedness between the actor and the recipient, B is the reproductive benefit provided to the recipient by the actor's behavior, and C is the reproductive cost to the actor for providing benefits to the recipient (Frank 1998:23). This rule can be reinterpreted in the human social context such that r is the psychological social distance between actor and recipient, C and B are measured in energetic terms, and frequencies of cooperative behaviors change due to human choice instead of natural selection. In this context, the proper measure of genotypic distance between two people is not their genetic relatedness but their cultural relatedness, captured here by the concept of psychological social distance. And psychological social distance is influenced by conceptual metaphors that govern social relations. These in turn promote or discourage the growth of social networks and all that comes with them.

In fact, one can take the analysis a step further by connecting the relationship between psychological social distance and altruism with key parameters of settlement scaling theory. According to the

~240~

latter, the area taken up by baseline productive units in a human settlement is given by

$$a = \left(\frac{\hat{g} a_0 l}{\varepsilon} \right)^{\alpha},$$

where $\hat{g}$ represents the average energetic benefit of interaction across all modes, ε the energetic cost of moving around, a_0 the distance at which interaction occurs, l the length of typical daily paths, and $2/3 \leq \alpha \leq 5/6$ (Bettencourt 2013; Ortman et al. 2014, 2015). The parameters $\hat{g}$ and ε may be interpretable as B and C from Hamilton's rule, respectively, in which case one can rewrite the rule as $r > \varepsilon/\hat{g}$. In other words, as the psychological social distance between individuals decreases (i.e., as r increases), people become willing to create social ties with people who are more physically and experientially distant. As a result, larger numbers of people can live in closer proximity and obtain increasing fractions of their needs through social interaction as opposed to from direct production. Thus, as psychological social distance decreases, organic solidarity increases.

Tracking Evolution in Cultural Genotypes

The discussion in this chapter makes clear that, in order to incorporate contemporary evolutionary theory into the study of social complexity, it is necessary to track evolution in the conceptual systems of social groups as well as their material behaviors. This is no small task because direct evidence for the initial emergence of social complexity comes primarily through archaeology. Fortunately, there are a range of techniques one can use to track cultural genotypes in addition to their material, phenotypic effects. In the final section of this chapter I discuss these methods in the hope of guiding initial research in this area.

The first approach one can utilize to study the effect of cultural genotypes on social behavior is social psychology experiments that examine how metaphors of physical distance and of kinship influence psychological social distance. Psychological distance is the sense of how close or far objects, events, and people are from the present self (Lewin 1943, 1951). In the domain of *temporal distance*, there is substantial and mounting evidence that changes in psychological distance profoundly influence a host of

Just as psychological temporal distance can influence people's reactions to other times, psychological social distance can influence people's reactions to other people.

outcomes: people are more willing to forgive past transgressions when those transgressions are psychologically distant than when they are psychologically close (Wohl and McGrath 2007); they are more motivated to prepare for upcoming examinations that are psychologically close than those that are psychologically far away (Peetz et al. 2009); and they glean greater enjoyment from daily life when a looming life transition such as college graduation is psychologically close (Kurtz 2008). Psychological distance also explains people's discounting of future outcomes independent of their objective temporal distance (Kim and Zauberman 2009). And changes in psychological distance can reduce the gap between people's intentions and their behaviors (Peetz et al. 2010).

Just as psychological temporal distance can influence people's reactions to other times, psychological social distance can influence people's reactions to other people (Van Boven et al. 2010; Williams and Bargh 2008). For example, people experience heighted empathy when there is overlap between their sense of self and others (Batson et al. 1997), and such empathic concern increases the utility people glean from others' outcomes. Such social utility effectively changes the structure of economic exchanges such that social dilemmas, typified by the oft-examined prisoner's dilemma, become problems of coordination, where the challenge is for individuals to mutually settle upon coordination rather than competition (Gibbons and Van Boven 2001; Rabin 1993). Reducing psychological social distance between people thus facilitates economic exchange in increasingly large social environments. But what influences psychological social distance?

The framework developed in this chapter suggests psychological social distance is often conceptualized via metaphors of kinship (Van Boven and Caruso 2015; Williams and Bargh 2008). ~243~ Kinship is strongly associated with subjective closeness (Neyer and Lang 2003), so metaphors that activate the experiential and emotional correlates of kinship should act to reduce psychological social distance. This relationship can be examined through laboratory and field experiments in which people are exposed to kinship metaphors. For example, one can prime people with words related to kinship, expose people to narratives that emphasize kinship, and ask people to engage in embodied movements that express kinship, such as ceremonies in which individuals are welcomed into a metaphorical family. The framework developed here predicts that such procedures will reduce the psychological social distance to nonkin individuals and lead to increased trust and cooperation in simple economic games.

Psychological distance is also often conceptualized in terms of physical distance (Williams and Bargh 2008). Spatial relationships are among the earliest concepts acquired in human development (Leslie 1982; Mandler 1992) and serve as the scaffolding upon which many abstract concepts are built (Boroditsky 2000; Gentner 2001; Williams and Bargh 2008). Specifically, mental representations of direct experience of spatial distance are often recruited to conceptualize psychological distance. For example, the idea that travel through time has a direction—the so-called "arrow of time"—implies a spatial relationship (Casasanto et al. 2010). Spatial units often define temporal units, as when "one year" is defined as the distance the earth travels in one revolution around the sun. People's mental representations of psychological social distance are therefore grounded in their mental representations of space. Spatial distance metaphors provide the language of psychological distance.

One can examine this conceptualization of psychological social distance by exposing people to metaphors of spatial proximity, such as having them locate points that are proximal rather than distal in geometric space (Williams and Bargh 2008). The expectation would be that exposure to proximal spatial distance will reduce psychological social distance and increase trust and cooperation in economic games. In other words, one would expect spatial metaphors to shape psychological distance and economic exchange in much the same way as kinship metaphors.

The second major method for tracking cultural genotypes related to social coordination involves identification of social distance metaphors in archaeological material culture and in historical linguistic records. A variety of established archaeological methods can be used to track the behavioral and material correlates of cultural genotypes, including settlement pattern analysis (Blanton et al. 1993; Glowacki and Ortman 2012; G. A. Johnson 1977, 1987), scaling analysis (Ortman et al. 2014, 2015), demographic reconstruction (Kohler and Reese 2014; Ortman 2014; J. Parsons 1971; Varien et al. 2007), analysis of production and exchange (Blakeslee 2012; Costin 1991; Hodder and Orton 1976; Leonard 2006; Renfrew 1984), and accumulations research (Jongman 2014; Potter and Ortman 2004; Varien and Mills 1997; Varien and Ortman 2005). However, the tracing of cultural genotypes is more challenging because, as mentioned above, the most relevant genotypes for the study of social

TABLE 2 Mapping of image-schematic correspondences between gardens and villages (after Ortman 2011a:Table 5.7).

Cobble-bordered gardens	Pueblo village
Image-schematic correspondences	
Rectangular shape of cells	Rectangular shape of rooms/plazas
Corn germinates underground	Children grow in houses
Corn plants grow in cells	Women work in the village
Water comes from mountains	Men work outside the village
Corn planted in old houses	People live in current houses
Water follows arroyos to gardens	Men follow trails to village
Cobs grow on plants in cells	Corn changed to food in plazas
Water causes kernels to germinate	Men cause babies to grow
Additional transferred structure	
Grid cell size	Room size
Cobble border	Room foundation
Upright cobbles mark corners	Upright cobbles mark corners
Earth within cell	Adobe house
Kernels planted in earth within cells	Children buried in plazas/houses
Water comes from clouds	Elders buried in ash piles
Clouds form over mountains	Men pray at mountain shrines
Cobs grow on plants in cell	Corn changed to food in plaza
Water germinates seeds	Cloud beings mingle with corn maidens in dances

~245~

complexity were first invented in societies that did not leave written records. Fortunately, research has shown that conceptual metaphors are often fossilized in language (Campbell 1998:254–273; Pinker 2007:235–278; Sweetser 1990:23–48; Traugott 1989) and are systematically expressed in material culture (Hays-Gilpin 2008; Ortman 2011a; Preston Blier 1987; Sekaquaptewa and Washburn 2004; Tilley 1999; Walens 1981; Whitley 2008; Whittlesey 2009; Wiseman 2014). As a result, the evolution of conceptual metaphors can be studied linguistically through etymology, polysemy, and semantic change among cognates across related languages (Ortman 2011a, 2011b, 2012; Schoenbrun 2012), and empirical generalizations on the ways humans express conceptual metaphors in everyday speech can be used to identify and trace the evolution of conceptual metaphors in material culture (Ortman 2000, 2008, 2012; Ortman and Bradley 2002).

It may prove helpful to provide an example. In my own research I have utilized the methods described above to clarify the role of cultural genotypes in the emergence of ancestral Tewa society in the

---

Archaeolinguistic methods provide evidence for the specific conceptual metaphors behind changing material expressions.

---

thirteenth century CE. This society took shape following a period of widespread social unrest that led to the migration of several tens of thousands of people from the Mesa Verde region of southwest Colorado and southeast Utah to the Northern Rio Grande region in New Mexico (Ortman 2010, 2012, 2014). In subsequent centuries these immigrants created a society that contained larger, more peaceful, and more prosperous communities than had ever existed before: community populations increased from a maximum of about 500 people to about 3,000 people (Crown et al. 1996; Glowacki and Ortman 2012; Varien et al. 1996); rates of interpersonal violence declined significantly (Kohler et al. 2014); community-scale economic specialization became widespread (Nelson and Habicht-Mauche 2006); and households maintained larger inventories of pottery and other goods (Ortman 2016).

Several lines of evidence suggest that this transformation involved the replacement of cultural genotypes that had supported communities of competitive kin groups by new genotypes that promoted cooperative, place-based communities. For example, Mesa Verde villages were collections of houses inhabited by groups of relatives, each with its own ceremonial structure, living space and storage space (Bradley 1993; Hegmon et al. 2000; Lipe 1989; Rohn 1965). These groups became increasingly unequal during the thirteenth century as some communities were forced to occupy less-productive and less-reliable land (Glowacki and Ortman 2012:Figure 14.8), as domestic architecture became increasingly diverse and elaborate (Lipe 2002), and as certain families came to control community surpluses (Lipe and Ortman 2000). In contrast, Tewa Basin settlements that formed in the aftermath of migration contain apartment-style dwellings that were built as a unit surrounding central plazas and sharing a single ceremonial structure or kiva (Creamer 1993; Kohler and Root 2004; Ruscavage-Barz and Bagwell 2006; Van Zandt 1999). All traces of lineage-based distinction had been erased.

~247~

The same pattern is apparent in other areas. As examples: thirteenth-century Mesa Verde domestic architecture contained elaborate kivas and towers, but these elaborations were stripped away

---

It is possible to track the evolution of cultural genotypes using methods grounded in historical linguistics and CMT.

---

from ancestral Tewa architecture in favor of the simpler forms of the Rio Grande tradition (Lakatos 2007). Also, the relatively baroque pottery assemblages of Mesa Verde households were simplified dramatically, with many functionally specialized vessel forms essentially dropping out of the record (Ortman 2012:Chapter 13). Finally, typical household organization changed from an extended-family pattern to a nuclear-family pattern based on the number of corn grinding bins found together in houses (Hegmon et al. 2000; Ortman 1998).

Archaeolinguistic methods provide evidence for the specific conceptual metaphors behind these changing material expressions.

For example, one of the ways Mesa Verde people imagined the community was as a group of relatives who ate from a communal bowl. This is reflected in the architecture of thirteenth-century Mesa Verde region villages and in the semantic history of *bu'u*, a term for plaza, village, and "large, low roundish place" that derives from an old word for "pottery bowl" (Ortman and Bradley 2002). Today, in contrast, Tewa people typically describe the community as a garden, with the women as corn and the men as clouds. This is reflected in the etymology of *owîngeh*, the dominant contemporary term for "village," and by the lyrics of songs composed for community ceremonies (Ortman 2011a). Table 2 summarizes the image-schematic mapping represented by this concept. Place-names and place lore surrounding ancestral Tewa settlements suggest THE PUEBLO IS A GARDEN replaced the older, container-based metaphor during the migration period (Ortman 2010).

~248~

The semantics of kin terms appear to have changed in parallel ways at this time. Kin terms in Tewa and related languages (Dozier 1955; E. Parsons 1932; Trager 1943) suggest that the Tanoan-speaking ancestors of Tewa people had separate terms for maternal and paternal grandmothers, and separate terms for maternal aunts, maternal uncles, paternal aunts, and paternal uncles. These early Tanoan speakers also appear to have distinguished between older and younger siblings. This suggests a bilateral age-graded kinship system in which multigenerational households were common. However, at some point after Tewa became distinct from Proto-Tiwa (ca. 1000 CE) but before the divergence of Arizona Tewa from Rio Grande Tewa (1700 CE), the meanings of these terms changed dramatically. The relative ages of siblings continued to be distinguished but not their gender, maternal vs. paternal relatives came to be referred to using a single term, and a number of terms came to be mapped onto all members of the pueblo. For example, any person of one's grandparents' generation came to be called "mother" or "father," and anyone of one's parents' generation came to be called "aunt" or "uncle" (see Dozier 1955). These changes reflect the erasure of bilateral lineage from the kinship classification and the extension of family relations to the entire community. In other words, these changes in the semantics of Tanoan kin terms imply that the cultural genotype THE PUEBLO IS A FAMILY came to be an important metaphor that promoted social coordination among people who were not actual relatives but who lived in the same settlement. These changes mirror the expanded social scale of the kiva from a ceremonial space for individual families to a space utilized by the entire community.

~249~

Finally, Tewa oral tradition suggests that new political institutions that enforced these emerging values were also established during the migration period. In contemporary Tewa communities the *K'ósa*, or "clowns," are important enforcers of social cohesion. The clowns are viewed as being *tepíngéh*, "of the middle of the house," meaning that they mediate between the social factions that comprise a Tewa village in a number of contexts. During public dances, the *K'ósa* typically meander among the dancers, pantomiming their motions, and singing out of time and tune. They also walk around the community telling jokes about members who have not been behaving properly in an attempt to shame them into better behavior. Tewa oral tradition consistently indicates that the *K'ósa* society was established during the migration period. For example, in the early twentieth century, Edward Curtis (1926:18) was told by a Tewa elder that "the *K'ósa* 'come from the north,' that is, the society was instituted in the ancient home of the Tewa." Harrington (1916:564) also recorded a statement that "the *K'ósa*, a mythic person who founded the *K'ósa* society," first appeared to Tewa ancestors when they lived in the Montezuma Valley. Finally, Tewa origin narratives themselves state that the *K'ósa* society was one of the societies established in preparation for the migration of Tewa ancestors from their ancestral home to the Tewa Basin (Ortiz 1969:15).

This example illustrates that it is possible to track the evolution of cultural genotypes using methods grounded in historical linguistics and CMT. Although it is difficult to track changes in cultural genotype frequencies using historical linguistic methods, this can be done when one has a corpus of dated written texts to work with, or when specific metaphors have concrete expressions in archaeological material culture (Ortman 2000, 2008). Thus, methods for implementing the research program on cultural genotypes and social complexity suggested in this chapter are not beyond our grasp.

Summary

Cultural evolutionary theory has not had much impact on the study of social complexity due to limitations of the neo-Darwinian view of evolution. However, new perspectives emerging from evolutionary biology—including an expanded notion of the genotype, the concept of genotype-phenotype maps, and the concept of hierarchical information—create the space for a reconceptualization of the cultural evolutionary process that is much more compatible with the empirical record of emergent social complexity. I have attempted to provide an initial reformulation that leads to the hypothesis that the ultimate driver of social complexity is the invention and spread of cultural genotypes that promote increases in the scale of social coordination. These cultural genotypes take the form of conceptual metaphors that map the bodily (and thus emotional) experience of closeness onto relationships with experientially distant persons. These mappings—often utilizing the domains of the family and of physical proximity—reduce the psychological social distance between individuals and transform innate prosocial instincts for family and friends into increasingly abstract social norms and values. As a result, more people can live in close proximity and interact regularly, and the many benefits of scale—including increases in productivity, the division of labor, economic exchange, knowledge generation and storage, and institutional development—can all occur.

I have also outlined a program of research on cultural genotypes and social complexity that has two major components. The first is social psychology experiments that gauge the effects of conceptual metaphors that utilize kinship and physical distance as the source domain to conceptualize the target domain of psychological ~251~ social distance, and thus socioeconomic interaction. The second component is integrated archaeolinguistic research that identifies and tracks the evolution of conceptual systems through expressions of their constituent metaphors in material culture and in language, and quantification of material properties of the social networks associated with these changing concepts through a variety of traditional archaeological methods. Previous research on changing cultural genotypes and behavioral patterns associated with Tewa Pueblo origins shows that this can be done. My hope is that this approach might be applied in other settings and with respect to societies organized at a variety of scales, and that the results of such studies will lead to a stronger role for evolutionary theory in the study of social complexity. ⸙

ACKNOWLEDGEMENTS

I wish to thank Leaf Van Boven, Doug Erwin, Andreas Wagner, David Schoenbrun, Anne Kandler, and Luís Bettencourt for discussions related to various elements of the framework developed in this chapter.

REFERENCES CITED

Acemoglu, Daron, and James A. Robinson
2012   *Why Nations Fail: The Origins of Power, Prosperity and Poverty.* Crown Business, New York.

Angel, Schlomo, Jason Parent, Daniel L. Civco, Alexander Blei, and David Potere
2011   The Dimensions of Global Urban Expansion: Estimates and Projections for All Countries, 2000–2050. *Progress in Planning* 75:53107.

Arthur, W. Brian
2009   *The Nature of Technology: What It Is and How It Evolves.* Free Press, New York.

Batson, C. Daniel, Karen Sager, Eric Garst, and Misook Kang
1997   Is Empathy-Induced Helping Due to Self-Other Merging? *Journal of Personality and Social Psychology* (73):495–509.

Batty, M.
2008   The Size, Scale, and Shape of Cities. *Science* 319:769–771.

Bettencourt, Luis M. A.
2013   The Origins of Scaling in Cities. *Science* 340:1438–1441.
2014   Impact of Changing Technology on the Evolution of Complex Informational Networks. *Proceedings of the IEEE* 102(12):1878–1891.

Bettencourt, Luis M. A., J. Lobo, D. Helbing, C. Kühnert, and G. B. West
2007   Growth, Innovation, Scaling, and the Pace of Life of Cities. *Proceedings of the National Academy of Science of the U.S.A.* 104:7301–7306.

Bettencourt, Luis M. A., J. Lobo, and D. Strumsky
2007   Invention in the City: Increasing Returns to Patenting as a Scaling Function of Metropolitan Size. *Research Policy* 36:107–120.

Bettencourt, Luis M. A., Horacio Samaniego, and HyeJin Youn
2014   Professional Diversity and the Productivity of Cities. *Scientific Reports* 4:5393. DOI:5310.1038/srep05393.

Blakeslee, Donald J.
2012   The Windom Pipe: A Chaine Operatoire Analysis. *Plains Anthropologist* 57(224):299–323.

Blanton, Richard E., and Lane Fargher
2008   *Collective Action in the Formation of Pre-Modern States.* Springer, New York.

Blanton, Richard E., Stephan A. Kowalewski, Gary Feinman, and Jill
Appel
1993   *Ancient Mesoamerica: A Comparison of Change in Three
Regions.* 2nd ed. Cambridge University Press, Cambridge.

Boroditsky, Lera
2000   Metaphoric Structuring: Understanding Time Through
Spatial Metaphors. *Cognition* 75:1–28.

Bowles, Samuel, and Herbert Gintis
2011   *A Cooperative Species: Human Reciprocity and Its Evolution.*
Princeton University Press, Princeton, New Jersey.

Boyd, Robert, and Peter J. Richerson
1985   *Culture and the Evolutionary Process.* University of Chicago   ~253~
Press, Chicago.

Bradley, Bruce A.
1993   Planning, Growth, and Functional Differentiation at a
Prehistoric Pueblo: A Case Study from SW Colorado. *Journal of Field
Archaeology* 20:23–42.

Campbell, Lyle
1998   *Historical Linguistics: An Introduction.* MIT Press,
Cambridge, Massachusetts.

Carneiro, Robert L.
1962   Scale Analysis as an Instrument for the Study of Cultural
Evolution. *Southwestern Journal of Anthropology* 18:149–169.
1967   On the Relationship Between Size of Population and
Complexity of Social Organization. *Southwestern Journal of
Anthropology* 23:234–243.
2000   The Transition from Quantity to Quality: A Neglected
Causal Mechanism in Accounting for Social Evolution. *Proceedings of
the National Academy of Science of the U.S.A.* 97(23):12926–12931.

Carroll, Sean B.
2006   *Endless Forms Most Beautiful: The New Science of Evo-Devo.*
W. W. Norton, New York.

Casasanto, Daniel, Olga Fotakopoulou, and Lera Boroditsky
2010   Space and Time in the Child's Mind: Evidence for a Cross-
Dimensional Asymmetry. *Cognitive Science* 34(3):387–405.

Cavalli-Sforza, Luigi Luca, and Marcus W. Feldman
1981   *Cultural Transmission and Evolution: A Quantitative
Approach.* Princeton University Press, Princeton, New Jersey.

Chan, Cindy, Leaf Van Boven, Eduardo B Andrade, and Dan Ariely
2014   Moral Violations Reduce Oral Consumption. *Journal of
Consumer Psychology* 24(3):381–386.

Costin, Cathy Lynn
   1991   Craft Specialization: Issues in Defining, Documenting, and
   Explaining the Organization of Production. In *Archaeological Method
   and Theory, vol. 3*, edited by Michael B. Schiffer, pp. 1–56. University
   of Arizona Press, Tucson.

Creamer, Winifred
   1993   *The Architecture of Arroyo Hondo Pueblo, New Mexico.*
   Arroyo Hondo Archaeological Series 7. School of American Research
   Press, Santa Fe, New Mexico.

Crown, Patricia L., Janet D. Orcutt, and Timothy A. Kohler
   1996   Pueblo Cultures in Transition: The Northern Rio Grande.
   In *The Prehistoric Pueblo World, A.D. 1150–1350*, edited by Michael
   A. Adler, pp. 188–204. University of Arizona Press, Tucson.

Curtis, Edward S.
   1926   *The North American Indian.* Vol. 17. J. P. Morgan, New
   York.

Damasio, Antonio
   1994   *Descartes' Error: Emotion, Reason, and the Human Brain.* G.
   P. Putnam, New York.

Dawkins, Richard
   1986   *The Blind Watchmaker: Why the Evidence of Evolution
   Reveals a Universe Without Design.* Norton, New York.

Dozier, Edward P.
   1955   Kinship and Linguistic Change Among the Arizona Tewa.
   International Journal of American Linguistics 21(3):242–257.

Durham, William H.
   1991   *Coevolution: Genes, Culture, and Human Diversity.* Stanford
   University Press, Stanford, California.

Durkheim, Emile
   1968   *The Elementary Forms of the Religious Life.* Free Press, New
   York. Originally published 1915, George Allen and Unwin.

Erwin, Douglas H.
   2012   Novelties that Change Carrying Capacity. *Journal of
   Experimental Zoology (Molecular and Developmental Evolution)*
   318B:460–465.

Erwin, Douglas H., and Eric H. Davidson
   2009   The Evolution of Hierarchical Gene Regulatory Networks.
   Nature Reviews Genetics 10(2):141–148.

Erwin, Douglas H., and James W. Valentine
2013   *The Cambrian Explosion: The Construction of Animal Biodiversity.* Roberts and Company, Greenwood Village, Colorado.

Fauconnier, Gil
1997   *Mappings in Thought and Language.* Cambridge University Press, Cambridge.

Fauconnier, Gil, and Mark Turner
1994   *Conceptual Projection and Middle Spaces.* Report 9401, UCSD Department of Cognitive Science, La Jolla, California.
2002   *The Way We Think: Conceptual Blending and the Mind's Hidden Complexities.* Basic Books, New York.

Feinman, Gary   ~255~
2011   Size, Complexity and Organizational Variation: A Comparative Approach. *Cross-Cultural Research* 45(1):37–59.

Feinman, Gary M., and Joyce Marcus (editors)
1998   *Archaic States.* School of American Research Press, Santa Fe, New Mexico.

Feldman, Jerome A.
2006   *From Molecule to Metaphor: A Neural Theory of Language.* MIT Press, Cambridge, Massachusetts.

Ferrada, Evandro
2014   The Amino Acid Alphabet and the Architecture of the Protein Sequence-Structure Map. I. Binary Alphabets. *PLoS Computational Biology* 10(12):e1003946.

Flannery, Kent, and Joyce Marcus
2012   *The Creation of Inequality: How Our Prehistoric Ancestors Set the Stage for Monarchy, Slavery, and Empire.* Harvard University Press, Cambridge, Massachusetts.

Ford, Richard I.
1972   An Ecological Perspective on the Eastern Pueblos. In *New Perspectives on the Pueblos*, edited by Alfonso Ortiz, pp. 1–17. University of New Mexico Press, Albuquerque.

Frank, Steven A.
1998   *Foundations of Social Evolution.* Princeton University Press, Princeton, New Jersey.

Gell-Mann, Murray
2011   Regularities in Human Affairs. *Cliodynamics* 2(1):52–70.

Gentner, Dedre
2001   Spatial Metaphors in Temporal Reasoning. In Spatial
Schemas and Abstract Thought, edited by M. Gattis, pp. 203–222.
MIT Press, Cambridge, Massachusetts.

Gibbons, Robert, and Leaf Van Boven
2001   Contingent Social Utility in the Prisoners' Dilemma.
Journal of Economic Behavior and Organization 45:1–17.

Glaeser, E. L., and B. Sacerdote
1999   Why Is There More Crime in Cities? *Journal of Political
Economy* 107:S225–S258.

Glowacki, Donna M., and Scott G. Ortman
2012   Characterizing Community-Center (Village) Formation
in the VEP Study Area. In *Emergence and Collapse of Early Villages:
Models of Central Mesa Verde Archaeology*, edited by Timothy A.
Kohler and Mark D. Varien, pp. 219–246. University of California
Press, Berkeley.

Gould, Stephen J.
1989   *Wonderful Life: The Burgess Shale and the Nature of
History*. W. W. Norton, New York.

Haidt, Jonathan
2012   *The Righteous Mind: Why Good People Are Divided by
Politics and Religion*. Pantheon Books, New York.

Harrington, John Peabody
1916   The Ethnogeography of the Tewa Indians. In *29th Annual
Report of the Bureau of American Ethnology*, pp. 29–618. Government
Printing Office, Washington, DC.

Hays-Gilpin, Kelley A.
2008   Life's Pathways: Geographic Metaphors in Ancestral
Puebloan Material Culture. In *Archaeology Without Borders: Contact,
Commerce and Change in the U.S. Southwest and Northwestern
Mexico*, edited by Laurie Webster and Maxine McBrinn, pp. 257–270.
University Press of Colorado, Boulder.

Hegmon, Michelle
1989   Social Integration and Architecture. In *The Architecture of
Social Integration in Prehistoric Pueblos*, edited by William D. Lipe and
Michelle Hegmon, pp. 5–14. Occasional Papers, no. 1. Crow Canyon
Archaeological Center, Cortez, Colorado.

Hegmon, Michelle, Scott G. Ortman, and Jeannette L. Mobley-Tanaka
    2000   Women, Men, and the Organization of Space. In *Women and Men in the Prehispanic Southwest: Labor, Power, and Prestige,* edited by Patricia L. Crown, pp. 43–90. School for American Research Press, Santa Fe, New Mexico.

Henrich, Joseph
    2001   Cultural Transmission and the Diffusion of innovations: Adoption Dynamics Indicate That Biased Cultural Transmission Is the Predominate Force in Behavioral Change. *American Anthropologist* 103:992–1013.
    2004   Demography and Cultural Evolution: How Adaptive Cultural Processes Can Produce Maladaptive Losses: The Tasmanian Case. *American Antiquity* 69(2):197–214.

Hill, Kim R., Brian M. Wood, Jacopo Baggio, A. Magdalena Hurtado, and Robert T. Boyd
    2014   Hunter-Gatherer Inter-Band Interaction Rates: Implications for Cumulative Culture. *PLoS* ONE 9(7):e102806.

Hodder, Ian
    1990   *The Domestication of Europe: Structure and Contingency in European Societies.* Blackwell, Oxford.

Hodder, Ian, and Clive Orton
    1976   *Spatial Analysis in Archaeology.* Cambridge University Press, Cambridge.

Hofstadter, Douglas, and Emmanuel Sander
    2013   *Surfaces and Essences: Analogy as the Fuel and Fire of Thinking.* Basic Books, New York.

Johnson, Gregory A.
    1977   Aspects of Regional Analysis in Archaeology. *Annual Review of Anthropology* 6:479–508.
    1987   The Changing Organization of Uruk Administration on the Susiana Plain. In *The Archaeology of Western Iran: Settlement and Society from Prehistory to the Islamic Conquest,* edited by Frank Hole, pp. 107–140. Smithsonian Institution Press, Washington, DC.

Johnson, Mark
    1987   *The Body in the Mind: The Bodily Basis of Meaning, Imagination, and Reason.* University of Chicago Press, Chicago.

Jongman, Willem M.
    2014   Re-Constructing the Roman Economy. In *Cambridge History of Capitalism, Volume I: From Ancient Origins to 1848,* edited by Larry Neal and Jeffrey G. Williamson, pp. 75–100. Cambridge University Press, Cambridge.

~257~

Kahneman, Daniel
2011   *Thinking, Fast and Slow*. Farrar, Straus and Giroux, New York.

Kim, B. Kyu, and Gal Zauberman
2009   Perception of Anticipatory Time in Temporal Discounting. *Journal of Neuroscience, Psychology, and Economics* 2(2):91–101.

Kline, Michelle A., and Robert Boyd
2010   Population Size Predicts Technological Complexity in Oceania. *Proceedings of the Royal Society B*. DOI:10.1098/rspb.2010.0452.

Kohler, Timothy A., Scott G. Ortman, Katie E. Grundtisch, Carly Fitzpatrick, and Sarah M. Cole
2014   The Better Angels of Their Nature: Declining Violence through Time among Prehispanic Farmers of the Pueblo Southwest. *American Antiquity* 79(3):444–464.

Kohler, Timothy A., and Kelsey M. Reese
2014   Long and Spatially Variable Neolithic Demographic Transition in the North American Southwest. *Proceedings of the National Academy of Science of the U.S.A.* 111(28):10101–10106.

Kohler, Timothy A., and Matthew J. Root
2004   The Late Coalition and Earliest Classic on the Pajarito Plateau (A.D. 1250–1375). In *Archaeology of Bandelier National Monument: Village Formation on the Pajarito Plateau, New Mexico*, edited by Timothy A. Kohler, pp. 173–214. University of New Mexico Press, Albuquerque.

Kövecses, Zoltán
2002   *Metaphor: A Practical Introduction*. Oxford University Press, New York.

Krakauer, David C.
2011   Darwinian Demons, Evolutionary Complexity, and Information Maximization. *Chaos* 21:037110-037111-037112.

Kurtz, J. L.
2008   Looking to the Future to Appreciate the Present: The Benefits of Perceived Temporal Scarcity. *Psychological Science* 19(12):1238–1241.

Lakatos, Steven
2007   Cultural Continuity and the Development of Integrative Architecture in the Northern Rio Grande Valley of New Mexico, A.D. 600–1200. *Kiva* 73(1):31–66.

Lakoff, George
1987   *Women, Fire, and Dangerous Things: What Categories Reveal About the Mind.* University of Chicago Press, Chicago.
1993   The Contemporary Theory of Metaphor. In *Metaphor and Thought*, edited by Andrew Ortony, pp. 202–251. 2nd ed. Cambridge University Press, Cambridge.

Lakoff, George, and Mark Johnson
1980   *Metaphors We Live By.* University of Chicago Press, Chicago.
1999   *Philosophy in the Flesh: The Embodied Mind and Its Challenge to Western Thought.* Basic Books, New York.

Lakoff, George, and Zoltán Kövecses
1987   The Cognitive Model of Anger Inherent in American English. In *Cultural Models in Language and Thought*, edited by Dorothy Holland and Naomi Quinn, pp. 195–211. Cambridge University Press, Cambridge.

Lakoff, George, and Rafael Núñez
2000   *Where Mathematics Comes From.* Basic Books, New York.

Lakoff, George, and Mark Turner
1989   *More Than Cool Reason: A Field Guide to Poetic Metaphor.* University of Chicago Press, Chicago.

Laland, Kevin N., and Gillian R. Brown
2002   *Sense and Nonsense: Evolutionary Perspectives on Human Behavior.* Oxford University Press, Oxford.

Landau, Mark J., Daniel Sullivan, and Jeff Greenberg
2009   Evidence That Self-Relevant Motives and Metaphoric Framing Interact to Influence Political and Social Attitudes. *Psychological Science* 20(11):1421–1427.

Leonard, Kathryn
2006   Directionality and Exclusivity of Plains-Pueblo Exchange during the Protohistoric Period, AD 1450–1700. In *The Social Life of Pots: Glaze Wares and Cultural Dynamics in the Southwest, AD 1250–1680*, edited by Judith A. Habicht-Mauche, Suzanne L. Eckert, and Deborah L. Huntley, pp. 232–252. University of Arizona Press, Tucson.

Leslie, Alan M
1982   The Perception of Causality in Infants. *Perception* 11(2):173–186.

Lewin, Kurt
    1943   Defining the "Field at a Given Time." *Psychological Review*
    50(3):292.
    1951   *Field Theory in Social Science.* Harper, New York.

Libby, Eric, and William C. Ratcliff
    2014   Ratcheting the Evolution of Multicellularity. *Science*
    326:426–427.

Lipe, William D.
    1989   Social Scale of Mesa Verde Anasazi Kivas. In *The
    Architecture of Social Integration in Prehistoric Pueblos*, edited by
    William D. Lipe and Michelle Hegmon, pp. 53–71. Occasional Papers,
    no. 1. Crow Canyon Archaeological Center, Cortez, Colorado.
    2002   Social Power in the Central Mesa Verde Region, A.D.
    1150–1290. In *Seeking the Center Place: Archaeology and Ancient
    Communities in the Mesa Verde Region*, edited by Mark D. Varien and
    Richard H. Wilshusen, pp. 203–232. University of Utah Press, Salt
    Lake City.

Lipe, William D., and Scott G. Ortman
    2000   Spatial Patterning in Northern San Juan Villages, A.D.
    1050–1300. *Kiva* 66(1):91–122.

Mandler, J. M.
    1992   How to Build a Baby II: Conceptual Primitives.
    Psychological Review 15:587–604.

Maynard Smith, John, and Eors Szathmary
    1999   *The Origins of Life: From the Birth of Life to the Origins of
    Language.* Oxford University Press, Oxford.

McNerney, James, J. Doyne Farmer, Sidney Redner, and Jessika E. Trancik
    2011   Role of Design Complexity in Technology Improvement.
    Proceedings of the National Academy of Science of the U.S.A.
    108(22):9008–9013.

Morris, Ian
    2010   *Why the West Rules—For Now.* Farrar, Straus and Giroux,
    New York.

Nagy, Bela, J. Doyne Farmer, Quan M. Bui, and Jessika E. Trancik
    2012   Statistical Basis for Predicting Technological Progress. *PLoS
    ONE* 8(2):e52669.

Nelson, Kit, and Judith A. Habicht-Mauche
   2006   Lead, Paint, and Pots: Rio Grande Intercommunity Dynamics from a Glaze Ware Perspective. In *The Social Life of Pots: Glaze Wares and Cultural Dynamics in the Southwest, AD 1250– 1680*, edited by Judith A. Habicht-Mauche, Suzanne L. Eckert, and Deborah L. Huntley, pp. 197–215. University of Arizona Press, Tucson.

Neyer, Franz J., and Frieder R. Lang
   2003   Blood Is Thicker Than Water: Kinship Orientation Across Adulthood. *Journal of Personality and Social Psychology* 84(2):310–321.

Nordbeck, S.
   1971   Urban Allometric Growth. *Geografiska Annaler* 53:54–67.

Norenzayan, Ara
   2013   *Big Gods: How Religion Transformed Cooperation and Conflict*. Princeton University Press, Princeton, New Jersey.

North, Douglass C., John Joseph Wallis, and Barry R. Weingast
   2009   *Violence and Social Orders: A Conceptual Framework for Interpreting Recorded Human History*. Cambridge University Press, Cambridge.

Nowak, M.
   2006   Five Rules for the Evolution of Cooperation. *Science* 314:1560–1563.

Odling-Smee, F. J., Kevin N. Laland, and Marcus W. Feldman
   2003   *Niche Construction: The Neglected Process in Evolution*. Princeton University Press, Princeton, New Jersey.

Ortiz, Alfonso
   1969   *The Tewa World: Space, Time, Being and Becoming in a Pueblo Society*. University of Chicago Press, Chicago.

Ortman, Scott G.

1998   Corn Grinding and Community Organization in the Pueblo Southwest, A.D. 1150–1550. In *Migration and Reorganization: The Pueblo IV Period in the American Southwest*, edited by Katherine A. Spielmann, pp. 165–192. Anthropological Research Papers, no. 51. Arizona State University, Tempe.

2000   Conceptual Metaphor in the Archaeological Record: Methods and an Example from the American Southwest. *American Antiquity* 65(4):613–645.

2008   Architectural Metaphor and Chacoan Influence in the Northern San Juan. In *Archaeology Without Borders: Contact, Commerce, and Change in the U.S. Southwest and Northwestern Mexico*, edited by Laurie Webster and Maxine McBrinn, pp. 227–255. Proceedings of the 2004 Southwest Symposium. University Press of Colorado, Boulder.

2010   Evidence of a Mesa Verde Homeland for the Tewa Pueblos. In *Leaving Mesa Verde: Peril and Change in the Thirteenth Century Southwest*, edited by Timothy A. Kohler, Mark D. Varien, and Aaron Wright, pp. 222–261. University of Arizona Press, Tucson.

2011a   Bowls to Gardens: A History of Tewa Community Metaphors. In *Religious Transformation in the Late Prehispanic Pueblo World*, edited by D. M. Glowacki and Scott Van Keuren, pp. 84–108. University of Arizona Press, Tucson.

2011b   Using Cognitive Semantics to Relate Mesa Verde Archaeology to Modern Pueblo Languages. In *Rethinking Anthropological Perspectives on Migration*, edited by Graciela S. Cabana and Jeffery J. Clark, pp. 111–146. University Press of Florida, Gainesville.

2012   *Winds from the North: Tewa Origins and Historical Anthropology*. University of Utah Press, Salt Lake City.

2014   Uniform Probability Density Analysis and Population History in the Northern Rio Grande. *Journal of Archaeological Method and Theory*. DOI:10.1007/s10816-014-9227-6.

2016   Human Securities and Tewa Origins. In *The Archaeology of Human Experience*, edited by Michelle Hegmon. American Anthropological Association, Washington, DC.

Ortman, Scott G., and Bruce A. Bradley

2002   Sand Canyon Pueblo: The Container in the Center. In *Seeking the Center Place: Archaeology and Ancient Communities in the Mesa Verde Region*, edited by Mark D. Varien and Richard H. Wilshusen, pp. 41–78. University of Utah Press, Salt Lake City.

Ortman, Scott G., Andrew H. F. Cabaniss, Jennie O. Sturm, and Luis M. A. Bettencourt
2014   The Pre-History of Urban Scaling. *PLoS* ONE 9(2):e87902. DOI:87910.81371/journal.pone.0087902.

Ortman, Scott G., Andrew Cabaniss, Jennie O. Sturm, and Luis M. A. Bettencourt
2015   Settlement Scaling and Increasing Returns in an Ancient Society. *Science Advances* 1e00066. DOI:10.1126/sciadv.00066.

Ortman, Scott G. and Grant D. Coffey
2017   Settlement Scaling in Middle-Range Societies. *American Antiquity* 82(4):662–682.

Palmer, Gary B.   ~263~
1996   *Toward a Theory of Cultural Linguistics.* University of Texas Press, Austin.

Parsons, Elsie C.
1932   The Kinship Nomenclature of the Pueblo Indians. *American Anthropologist* 34(3):377–389.

Parsons, Jeffrey R.
1971   *Prehistoric Settlement Patterns in the Texcoco Region, Mexico.* Memoirs, No. 3. Museum of Anthropology, University of Michigan, Ann Arbor.

Pauketat, Timothy R.
2007   *Chiefdoms and Other Archaeological Delusions.* AltaMira Press, Lanham, Maryland.

Peetz, Johanna, Roger Buehler, and Anne Wilson
2010   Planning for the Near and Distant Future: How Does Temporal Distance Affect Task Completion Predictions? *Journal of Experimental Social Psychology* 46(5):709–720.

Peetz, Johanna, Anne E. Wilson, and Erin J. Strahan
2009   So far Away: The Role of Subjective Temporal Distance to Future Goals in Motivation and Behavior. *Social Cognition* 27(4):475–495.

Peregrine, Peter N., Carol R. Ember, and Melvin Ember
2004   Universal Patterns in Cultural Evolution: An Empirical Analysis Using Guttman Scaling. *American Anthropologist* 106(1):145–149.

Pinker, Stephen
2007   *The Stuff of Thought: Language as a Window into Human Nature.* Penguin, New York.
2011   *The Better Angels of Our Nature: Why Violence Has Declined.* Penguin, New York.

Potter, James M., and Scott G. Ortman
2004   Community and Cuisine in the Prehispanic American Southwest. In *Identity, Feasting, and the Archaeology of the Greater Southwest*, edited by Barbara J. Mills, pp. 173–191. University Press of Colorado, Boulder.

Powell, Adam, Stephen Shennan, and Mark G. Thomas
2009   Late Pleistocene Demography and the Appearance of Modern Human Behavior. *Science* 324:1298–1301.

Preston Blier, Susan
1987   *The Anatomy of Architecture: Ontology and Metaphor in Batammaliba Architectural Expression*. University of Chicago Press, Chicago.

Rabin, Matthew
1993   Incorporating Fairness into Game Theory and Economics. *The American Economic Review* 83(5):1281–1302.

Rappaport, Roy A.
1968   *Pigs for the Ancestors: Ritual in the Ecology of a New Guinea People*. Yale University Press, New Haven.
1979   *Ecology, Meaning, and Religion*. North Atlantic Books, Richmond, California.

Renfrew, Colin
1984   *Approaches to Social Archaeology*. Harvard University Press, Cambridge, Massachusetts.

Richerson, Paul J., and Robert Boyd
2005   *Not by Genes Alone: How Culture Transformed Human Evolution*. University of Chicago Press, Chicago.

Rohn, Arthur H.
1965   Postulation of socio-Economic Groups from Archaeological Evidence. In *Contributions of the Wetherill Mesa Archaeological Project*, edited by D. Osborne, pp. 65–69. Memoir, no. 19. Society for American Archaeology, Washington, DC.

Ruscavage-Barz, Samantha, and Elizabeth Bagwell
2006   Gathering Spaces and Bounded Places: The Religious Significance of Plaza-Oriented Communities in the Northern Rio Grande, New Mexico. In *Religion in the Prehispanic Southwest*, edited by Christine VanPool, Todd L. VanPool ,and David A. Phillips, Jr., pp. 81–101. AltaMira Press, Lanham, Maryland.

Sahlins, Marshall D.
1972   *Stone Age Economics*. Aldine, Chicago.

Samaniego, H., and Melanie E. Moses
 2009 Cities as Organisms: Allometric Scaling of Urban Road Networks. *Journal of Transportation and Land Use* 1:2139.

Schläpfer, Markus, Luis M. A. Bettencourt, Sebastian Grauwin, Mathias Raschke, Rob Claxton, Zbigniew Smoreda, Geoffrey B. West, and Carlo Ratti
 2014 The Scaling of Human Interactions with City Size. *Journal of the Royal Society Interface* 11:20130789.

Schoenbrun, David Lee
 2012 Mixing, Moving, Making, Meaning: Possible Futures for the Distant Past. *African Archaeological Review* 29(1):293–317.

Sekaquaptewa, Emory, and Dorothy Washburn   ~265~
 2004 They Go Along Singing: Reconstructing the Past from Ritual Metaphors in Song and Image. *American Antiquity* 69(3):457–486.

Sharif, Azim F., and Ara Norenzayan
 2007 God Is Watching You: Priming God Concepts Increases Prosocial Behavior in an Anonymous Economic Game. *Psychological Science* 18(9):803–809.

Shennan, Stephen
 2002 *Genes, Memes and Human History: Darwinian Archaeology and Cultural Evolution.* Thames & Hudson, London.

Shore, Bradd
 1996 *Culture in Mind: Cognition, Culture, and the Problem of Meaning.* Oxford University Press, Oxford.

Shubin, Neil
 2009 *Your Inner Fish: A Journey into the 3.5 Billion History of the Human Body.* Pantheon Books, New York.

Simpson, Carl
 2011 How Many Levels Are There? How Insights from Evolutionary Transitions in Individuality Help Measure the Hierarchical Complexity of Life. In *The Major Transitions in Evolution Revisited*, edited by B. Calcott and K. Sterelny, pp. 199–226. MIT Press, Cambridge, Massachusetts.
 2012 The Evolutionary History of Division of Labour. *Proceedings of the Royal Society B* 279:116–121.

Singer, Peter
 1981 *The Expanding Circle: Ethics, Evolution and Moral Progress.* Princeton University Press, Princeton, New Jersey.

Slingerland, Edward
　　2004　Conceptual Metaphor Theory as Methodology for Comparative Religion. *Journal of the American Academy of Religion* 72(1):1–31.
　　2008　*What Science Offers The Humanities: Integrating Body and Culture*. Cambridge University Press, Cambridge.

Smith, Adam T.
　　2003　*The Political Landscape: Constellations of Authority in Early Complex Polities*. University of California Press, Berkeley.

Smith, Michael E. (editor)
　　2012　*The Comparative Archaeology of Complex Societies*. Cambridge University Press, Cambridge.

Smith, Monica L. (editor)
　　2003　*The Social Construction of Ancient Cities*. Smithsonian Books, Washington, DC, and London.

Sveikauskas, L.
　　1975　The Productivity of Cities. *Quarterly Journal of Economics* 89:393–413.

Sweetser, Eve
　　1990　*From Etymology to Pragmatics: Metaphorical and Cultural Aspects of Semantic Structure*. Cambridge University Press, Cambridge.

Tilley, Christopher
　　1999　*Metaphor and Material Culture*. Blackwell Publishers, Oxford and London.

Trager, George L.
　　1943　The Kinship and Status Terms of the Tiwa Languages. *American Anthropologist* 45(4):557–571.

Traugott, Elizabeth Closs
　　1989　On the Rise of Epistemic Meanings in English: An Example of Subjectification in Semantic Change. *Language* 65(1):31–55.

Travers, Mark W.
　　2015　Phenomenological Foundations of Psychological Distance, PhD dissertation, Department of Psychology and Neuroscience, University of Colorado Boulder.

Trigger, Bruce G.
　　2003　*Understanding Early Civilizations*. Cambridge University Press, Cambridge.

Ur, Jason
  2014   Households and the Emergence of Cities in Ancient
  Mesopotamia. *Cambridge Archaeological Journal* 24(2):249–268.

Van Boven, Leaf, and Eugene M Caruso
  forthcoming   The Phenomenological Foundations of Psychological
  Distance. *Social and Personality Psychology Compass.*

Van Boven, Leaf, and Eugene M Caruso
  2015   The Tripartite Foundations of Temporal Psychological
  Distance: Metaphors, Ecology, and Teleology. *Social and Personality
  Psychology Compass* 2015:1–13.

Van Boven, Leaf, Joanne Kane, A. Peter McGraw, and Jeannette Dale
  2010   Feeling Close: Emotional Intensity Reduces Perceived   ~267~
  Psychological Distance. *Journal of Personality and Social Psychology*
  98:872–885.

Van Zandt, Tineke
  1999   Architecture and Site Structure. In *The Bandelier
  Archaeological Survey*, edited by Robert P. Powers and Janet
  D. Orcutt, pp. 309–388. Intermountain Cultural Resources
  Management, Professional Paper No. 57. National Park Service, Santa
  Fe, New Mexico.

Varien, Mark D., William D. Lipe, Michael A. Adler, Ian M. Thompson,
  and Bruce A. Bradley
  1996   Southwestern Colorado and Southeastern Utah Settlement
  Patterns: A.D. 1100 to 1300. In *The Prehistoric Pueblo World, A.D.
  1150–1350*, edited by Michael A. Adler, pp. 86–113. University of
  Arizona Press, Tucson.

Varien, Mark D., and Barbara J. Mills
  1997   Accumulations Research: Problems and Prospects for
  Estimating Site Occupation Span. *Journal of Archaeological Method
  and Theory* 4:141–191.

Varien, Mark D., and Scott G. Ortman
  2005   Accumulations Research in the Southwest United States:
  Middle-Range Theory for Big-Picture Problems. *World Archaeology*
  37(1):132–155.

Varien, Mark D., Scott G. Ortman, Timothy A. Kohler, Donna M.
  Glowacki, and C. David Johnson
  2007   Historical Ecology in the Mesa Verde Region: Results from
  the Village Project. *American Antiquity* 72(2):273–299.

Wagner, Andreas
  2011   *The Origins of Evolutionary Innovations*. Oxford University
  Press, Oxford.

Wagner, Gunter
    2014   *Homology, Genes, and Evolutionary Innovation.* Princeton University Press, Princeton, New Jersey.

Walens, S.
    1981   *Feasting with Cannibals: An Essay in Kwakiutl Cosmology.* Princeton University Press, Princeton, New Jersey.

Walker, Robert S., and Kim R. Hill
    2014   Causes, Consequences, and Kin Bias of Human Group Fissions. *Human Nature* 25:465–475.

Whitley, David S.
    2008   Archaeological Evidence for Conceptual Metaphors as Enduring Knowledge Structures. *Time & Mind* 1(1):7–30.

Whittlesey, S. M.
    2009   Mountains, Mounds and Meaning: Metaphor in the Hohokam Cultural Landscape. In *The Archaeology of Meaningful Places*, edited by Brenda J. Bowser and Maria-Nieves Zedeno, pp. 73–89. University of Utah Press, Salt Lake City.

Williams, Lawrence E., and John H. Bargh
    2008   Keeping One's Distance: The Influence of Spatial Distance Cues on Affect and Evaluation. *Psychological Science* 19(3):302–308.

Wiseman, Rob
    2014   Social Distance in Hunter-Gatherer Settlement Sites: A Conceptual Metaphor in Material Culture. *Metaphor and Symbol* 29(2):129–143.

Wohl, M. J. A., and A. L. McGrath
    2007   The Perception of Time Heals All Wounds: Temporal Distance Affects Willingness to Forgive Following an Interpersonal Transgression. *Personality and Social Psychology Bulletin* 33(7):1023–1035.

Yoffee, Norman
    2005   *Myths of the Archaic State: Evolution of the Earliest Cities, States and Civilizations.* Cambridge University Press, Cambridge.

Youn, HyeJin, Deborah Strumsky, Luis M. A. Bettencourt, and Jose Lobo
    2015   Invention as a Combinatorial Process: Evidence from US Patents. *Journal of the Royal Society Interface* 12:20150272.

Zhong, Chen-Bo, and Katie Liljenquist
    2006   Washing Away Your Sins: Threatened Morality and Physical Cleansing. *Science* 313:1451–1452.

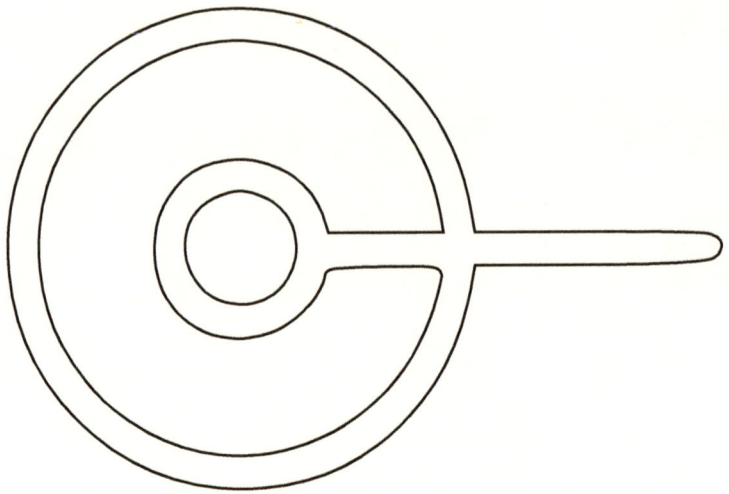

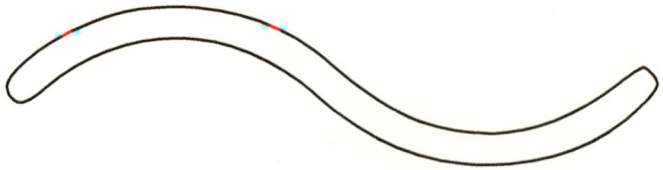

ꟼ

TOWARD A THEORY OF RECURRENT
SOCIAL FORMATIONS

Peter N. Peregrine, Lawrence University and Santa Fe Institute

Social formations with remarkable similarities in social, political, and economic structures have emerged repeatedly throughout history and in all areas of the globe. These recurrent social formations are thought to be the product of "general" evolutionary pro- ~271~ cesses (*sensu* Sahlins 1960) that act upon human societies regardless of time or place, and have been the focus of anthropological research since the very beginning of the field (e.g., Tylor 1871:1–14). An unfortunate consequence of early anthropologists' recognition of recurrent social forms was the linking of this observance with prevailing Eurocentric views of progress. Thus recurrent social formations were presented in the framework of a series of progressive "stages" through which all human societies progressed (or failed to progress) from, for example, "savagery" to "barbarism" to "civilization" (Morgan 1877).

The theory of recurrent social formations being the product of a steplike series of stages moving human societies toward an essentially European form was rejected by the turn of the twentieth century. Nothing replaced it until the 1960s, when anthropologists could no longer avoid the fact that an informal taxonomy of social types had emerged over the previous half-century which bore an uncanny resemblance to the stages proposed by earlier anthropologists (Sahlins 1960:40–44; Steward 1955:178–185). Concern shifted to creating a formal taxonomy of cultural types (Steward 1955:22–26; 87–92). This taxonomy took two forms. One focused on political economy: bands, tribes, chiefdoms, and states (Service 1962); the other focused on social stratification: egalitarian, ranked, and stratified (Fried 1967). These two taxonomies are effectively identical if one removes tribes as a type (i.e., bands are egalitarian, chiefdoms are ranked, and states are stratified).

Problems with these taxonomies have been recognized from the time they were put forward (e.g., Service 1962:182), yet they have become embedded in anthropological thinking to the extent that scholars cannot seem to work without them (Bailey 1994:1). By way of example I present a quote, though in doing so I do not intend to demean the author's work; the quote simply demonstrates how ingrained these cultural taxonomies have become:

~272~

> It is certainly true that terms such as chiefdom can cause misunderstandings because they are associated with baggage acquired as their generally accepted use has changed over time. Despite the potential for confusion, the society centered on Cahokia is referred to here as a chiefdom, more precisely a complex chiefdom. (Milner 1990:3)

Here a thoughtful scholar finds himself trapped into using a taxonomy he knows is flawed because he has no other conceptual framework for trying to explain to peers the similarities and differences between Cahokia and other social formations. And part of the problem in this specific case is that such explanations have historically degraded into a relatively unproductive taxonomic argument over whether Cahokia should be typed as a chiefdom or as a state. (In the quote above, the author attempts to avoid the argument by saying it is neither—it is a different type, a "complex chiefdom." Obviously, while this avoids the chiefdom versus state problem, it does not avoid the problem of being a taxonomic rather than explanatory argument.) The concept of chiefdoms as a cultural type has proven particularly thorny, as there is great variation among social formations defined as chiefdoms (Feinman and Neitzel 1984; Peebles and Kus 1977).

In recent years there has been much discussion about

developing taxonomic approaches based on sociopolitical processes rather than political-economic or social traits (e.g., Blanton
et al. 1996; Peregrine 2012). A problem with this approach is that
it is still essentially classificatory rather than explanatory. If this
approach is pursued, the taxonomic arguments do not drop away
but rather are transferred from traits to processes. The appeal is
that processes can be explanatory, but the approach does not frame
the effort in a directly explanatory manner.

What is needed, I suggest, is a broader theory to explain the
presence and persistence of recurrent social formations. This
theory would not in itself define those social formations but rather
would define processes out of which those recurrent formations
arise. In other words, the processes are not intrinsic to the social
formations themselves and are not what differentiate them from
one another; rather, the processes are epigenetic and autopoietic,
operating between and among social groups to shape them into
similar formations (*sensu* Bourdieu 1977:72).

As a way to begin, I consider three approaches to modeling evolutionary processes that provide both a means of clearly defining a
taxonomy and a general underlying theory to explain why such a
taxonomy might (or might not) exist. While these methods do not
translate directly into social theory, they do provide a clear pathway
toward developing such theory. These approaches to modeling evolution are Guttman scaling, morphospace analysis, and the exploration of adaptive landscapes.

Data Sources

It is important to understand the data sources to be used in the analyses that follow, as they are somewhat unconventional. The unit of
analysis here is the "archaeological tradition," defined as "a group
of populations sharing similar subsistence practices, technology,
and forms of socio-political organization, which are spatially

contiguous over a relatively large area and which endure temporally for a relatively long period" (Peregrine 2001:iv). Minimal area coverage for an archaeological tradition can be thought of as something like 100,000 km², while minimal temporal duration can be thought of as something like five centuries. However, these figures are meant to help clarify the concept of an archaeological tradition, not to formally restrict its definition to these conditions. Archaeological traditions are artificial units of analysis, but they are not arbitrary. They were designed to provide a unit that could be compared across broad regions and time scales. The sample used here includes all cases from the last 12,000 years listed in the *Outline of Archaeological Traditions* (Peregrine 2001), which is a comprehensive catalog of archaeological traditions covering the entire globe for the last two million years.

The coded data themselves come from two primary sources: the *Atlas of Cultural Evolution* (Peregrine 2003) and additional variables on societal scale coded for the *Atlas* cases by researchers at the Santa Fe Institute (these are discussed in more detail in Ortman et al., chapter 7 in this volume, and in Ortman et al. [2014]). The *Atlas of Cultural Evolution* provides basic data on the evolution of cultural complexity for all cases in *Outline of Archaeological Traditions* sample. Data for the *Atlas* (i.e., the data used in this paper) were coded from entries in the *Encyclopedia of Prehistory* (Peregrine and Ember 2001–2002), a nine-volume work providing summary information on all cases in the *Outline of Archaeological Traditions*. The base set of variables are 10 Likert-scale variables created by Murdock and Provost (1973) for their scale of cultural complexity. These 10 variables were recoded in two ways to better match the data available in the archaeological record. First, the individual traits measured through the five-item Likert scales were recoded into 15 present–absent variables (Table 1). Second, the five-item Likert scales were simplified into three-item

~274~

TABLE 1 15-item Murdock-Provost scale of cultural complexity, from Peregrine (2003). These are present-absent variables based upon key indicators of cultural complexity as defined by Murdock and Provost (1973) and were demonstrated to form a Guttman scale in the order presented here (Peregrine et al. 2004).

1. Ceramic production
2. Presence of domesticates
3. Sedentarism
4. Inegalitarian (status or wealth differences)
5. Density > 1 person/mi^2 specialists
6. Reliance on food production
7. Villages > 100 persons
8. Metal production
9. Social classes present
10. Towns > 400 persons population
11. State (3+ levels of hierarchy)
12. Density > 25 persons/mi^2
13. Wheeled transport
14. Writing of any kind
15. Money of any kind

~275~

scales (Table 2; see Peregrine [2003] for details of the recoding procedure).

Two additional scale variables were constructed for the analyses presented here. Chick (1997) conducted a factor analysis on data coded for the Standard Cross-Cultural Sample and found two factors underlying the Murdock and Provost scale of cultural complexity. One reflects the technological capabilities of the society while the other reflects the scale or size (in terms of population) of the society (Chick 1997). The variables that comprise each factor are shown in Table 3. These factor analysis–derived variables are employed here in the morphospace and adaptive landscape analyses. The recoded three-item Likert scale variables were used to derive the factor scores for each archaeological tradition.

TABLE 2 Murdock-Provost scale of cultural complexity as recoded by Peregrine (2003). The original five-item Likert scale variables were transformed into three-item scales to make them easier to code with archaeological data.

Scale 1: Writing and Records
  1 = None
  2 = Mnemonic or nonwritten records
  3 = True writing

Scale 2: Fixity of Residence
  1 = Nomadic
  2 = Seminomadic
  3 = Sedentary

Scale 3: Agriculture
  1 = None
  2 = 10% or more, but secondary
  3 = Primary

Scale 4: Urbanization (largest settlement)
  1 = Fewer than 100 persons
  2 = 100–399 persons
  3 = 400+ persons

Scale 5: Technological Specialization
  1 = None
  2 = Pottery
  3 = Metalwork (alloys, forging, casting)

Scale 6: Land Transport
  1 = Human only
  2 = Pack or draft animals
  3 = Vehicles

Scale 7: Money
  1 = None
  2 = Domestically usable articles
  3 = Currency

Scale 8: Density of Population
  1 = Less than 1 person/mi^2
  2 = 1–25 persons/mi^2
  3 = 26+ persons/mi^2

Scale 9: Political Integration
  1 = Autonomous local communities
  2 = 1 or 2 levels above community
  3 = 3 or more levels above community

Scale 10: Social Stratification
  1 = Egalitarian
  2 = 2 social classes
  3 = 3 or more social classes or castes

TABLE 3. Underlying factors in the Murdock-Provost scale of cultural complexity, as identified by Chick (1997). The variables comprising the two factors are listed in the order in which they loaded upon each factor. The technology factor explains 52.8 percent of the total variance while the scale factor explains 14.5 percent. Individual variable loadings are shown in parentheses.

Technology Factor	Scale Factor
Scale 1: Writing and Records (0.848)	Scale 2: Fixity of Residence (0.918)
Scale 6: Land Transport (0.846)	Scale 3: Agriculture (0.849)
Scale 10: Social Stratification (0.716)	Scale 8: Density of Population (0.824)
Scale 9: Political Integration (0.669)	Scale 4: Urbanization (0.542)
Scale 5: Technological Specialization (0.606)	
Scale 7: Money (0.578)	

Methods

GUTTMAN SCALING

Guttman scaling is a method for scaling items according to an underlying cumulative dimension (McIver and Carmines 1981:40). In a perfect Guttman scale, each item is cumulative in respect to the item below it on the scale; in other words, the presence of an item at the top of a Guttman scale indicates probabilistically that other traits lower on the scale are also present (Guttman 1950). A score of three on a given Guttman scale would, for example, indicate that items one, two, and three are all present, and that no other items in the scale are present.

Robert Carneiro (1962) suggested that Guttman scaling held great potential for the study of cultural evolution because it identifies a clear hierarchy among a group of scale items. There are obvious evolutionary implications if one finds that a group of traits form a Guttman scale, as "the order in which the traits are arranged, from bottom to top, is [probabilistically] the order in which the societies have evolved them" (Carneiro 1970:837; also see Gell-Mann 2011). Guttman scaling provides a method for modeling cultural evolution because the hierarchy of cultural traits inherent in a Guttman scale suggests an evolutionary order.

Peregrine, Ember, and Ember (2004) demonstrated that the 15-item version of the Murdock-Provost scale of cultural complexity forms a Guttman scale in the order with which the individual variables are presented in Table 1. According to this scale, the presence of status or wealth differences within a given society also indicates that the society is sedentary, has domesticated plants or animals, and produces ceramics. The implication of this scale is that there is at least one unidimensional evolutionary process at work to create this hierarchy.

Figure 1 presents scalograms based on this scale for eight regional cultural evolutionary sequences. The sequences provide an empirical picture of cultural change over time that supports the Guttman scale. But there is also an interesting pattern of jumps in which several scale items appear together. Indeed, the pattern appears to be fairly regular in several ways. First, there seem to be similar rapid leaps from societies having none of the traits to having agriculture and/or villages, implying that these traits appear together or in rapid succession. There appears a second common leap to a state form of government, with intervening traits appearing together or in train.

The steplike rather than smooth accumulation of traits suggests that the unidimensional process underlying the Guttman scale is not uniform in its effects. Rather, traits often appear in clusters or groups, an effect modeled in the dendrogram presented as Figure 2. It is interesting that these clusters of traits map onto existing typologies of recurrent social formations. Cluster A is similar to what are commonly called chiefdoms—sedentary,

FIGURE 1 Scalograms for eight regional cultural-evolutionary sequences, from Peregrine et al. (2004). Columns represent individual archaeological traditions from the Atlas of Cultural Evolution (Peregrine 2003) with their identification numbers at the bottom of the column. An X represents a match between the Guttman scale and the archaeological record; a ? represents the absence of an expected scale item.

Top scalogram

	4005	4015	4025	4030	4055	4060	5505	5515	5530	5535	5555	7040	6110	6115	6125	6130	6135	6205	6215	6220	6245	6250	6255	6260	6265	6270
Money										X	X					X	X				X	X	X	X	X	X
Writing						X				X	X					X	X				X	X	X	X	X	X
Wheel						X				X	X					X	X				X	X	X	X	X	X
Density >25						X			X	X	X					X	X				X	X	X	X	X	X
State						X			X	X	X					X	X				X	X	X	X	X	X
Towns >400					X	X			X	X	X					X	X				X	X	X	X	X	X
Classes					X	X			X	X	X					X	X				X	X	X	X	X	X
Metals					X	X			X	X	X					X	X				X	X	X	X	X	X
Villages >100			X	X	X	X		X	X	X	X					X	X	X	X	X	X	X	X	X	X	X
Agriculture		X	X	X	X	X		X	X	X	X			X	X	X	X	X	X	X	X	X	X	X	X	X
Density >1		X	X	X	X	X		X	X	X	X			X	X	X	X	X	X	X	X	X	X	X	X	X
Inegalitarian		X	X	X	X	X		X	X	X	X			X	X	X	X	X	X	X	X	X	X	X	X	X
Sedentarism		X	X	X	X	X		X	X	X	X			X	X	X	X	X	X	X	X	X	X	X	X	X
Domesticates		X	X	X	X	X		X	X	X	X			X	X	X	X	X	X	X	X	X	X	X	X	X
Ceramics		X	X	X	X	X		X	X	X	X			?	?	X	X	?	X	X	X	X	X	X	X	X

YELLOW RIVER VALLEY	INDUS RIVER VALLEY	NILE RIVER VALLEY	MESOPOTAMIA

~279~

Bottom scalogram

	6175	6180	6185	6190	2205	2215	2225	2235	2250	2260	2275	1505	1512	1522	1550	1560	1505	1510	1515	1520	1540	1570
Money																						
Writing															X	X						X
Wheel															?	?						?
Density >25															?	?						?
State				X				X	X	X	X				X	X				X	X	X
Towns >400				X			X	X	X	X	X				X	X				X	X	X
Classes				X			?	X	X	X	X				X	X				?	X	X
Metals				X			?	X	X	X	X				X	X				?	X	X
Villages >100			X	X		X	X	X	X	X	X			X	X	X			X	X	X	X
Agriculture		X	X	X		?	X	X	X	X	X			X	X	X			X	X	X	X
Density >1		?	X	X		?	X	X	X	X	X			X	X	X			X	X	X	X
Inegalitarian		X	X	X		?	X	X	X	X	X			X	X	X			X	X	X	X
Sedentarism		?	X	X		X	X	X	X	X	X			X	X	X			X	X	X	X
Domesticates		X	X	X		X	X	X	X	X	X			X	X	X			X	X	X	X
Ceramics		X	X	X		?	X	X	X	X	X		?	X	X	X		?	X	X	X	X

WEST AFRICA	HIGHLAND PERU	LOWLAND	HIGHLAND
		MESOAMERICA	

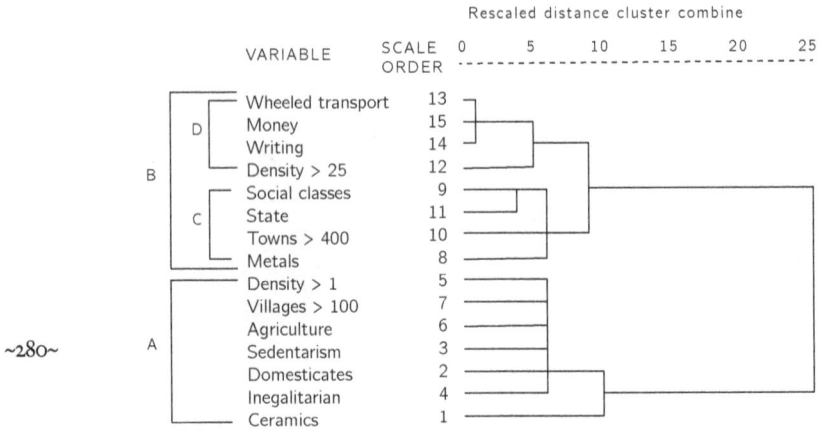

FIGURE 2 Ward's method dendrogram based upon the eight cultural evolutionary sequences presented in Figure 1, from Peregrine et al. (2004).

inegalitarian but non-state societies. Cluster B encompasses states, some large and bureaucratic (cluster D), some smaller and lacking scribes, money, and other elements of bureaucracy (cluster C).

The Guttman scale analyses thus suggest that recurrent social formations may be the result of a steplike or punctuated process in which a critical state is reached followed by a transformation, or, alternatively, that intermediate states are unstable. The transformed states are relatively stable and appear as recurrent social formations, although each evolves independently through the same transitive process. I suggest that what we see as recurrent social formations are not "stages" of development or societal "types" but rather are the results of an autopoietic process of convergent evolution acting across societies through time. I explore this idea further in the next section.

THEORETICAL AND EMPIRICAL MORPHOSPACES

One way to explore the boundaries or limits of convergent evolution is through the analysis of theoretical and empirical morphospaces. A morphospace is an n-dimensional space constructed from morphological variables for a given species or other taxon. A theoretical morphospace is produced using variables that are thought to be important in understanding the range of variation within a particular taxon. These variables make up the morphospace's dimensions and represent the complete range of possible forms along these dimensions (McGhee 2007:57–61). An empirical morphospace maps known cases onto the morphospace to identify the range of actual forms in existence. The absence of forms in a particular area of a morphospace suggests regions that are either impossible or unsuccessful (McGhee 2007:72–75).

~281~

Morphospace analysis was developed by Raup (1966), who used the method to demonstrate that species with coiled shells, because of the physical parameters of coiling, are limited in the range of possible morphospaces within which they can exist. This is considered an architectural or geometric limitation. Functional or adaptive limitations can also be found; for example, Chamberlain (1976) argued that the empirical range of ammonoid morphospace was limited because of the need for efficient swimming. Inefficient forms died out, and thus the empirical range of ammonoid forms is far smaller than the theoretical range. It is this ability to explore the functional or adaptive significance of the variation between empirical and theoretical morphospaces that makes morphospace analysis a particularly valuable tool for examining evolution (see also Atkinson and Whitehouse 2011).

Figure 3 presents a two-dimensional morphospace for the two variables derived by factor analysis from the Murdock and Provost scale of cultural complexity (Chick 1997). As presented in Figure 3, the technology factor is the y-axis and the scale factor is the x-axis. One would assume that in empirical morphospace societies would

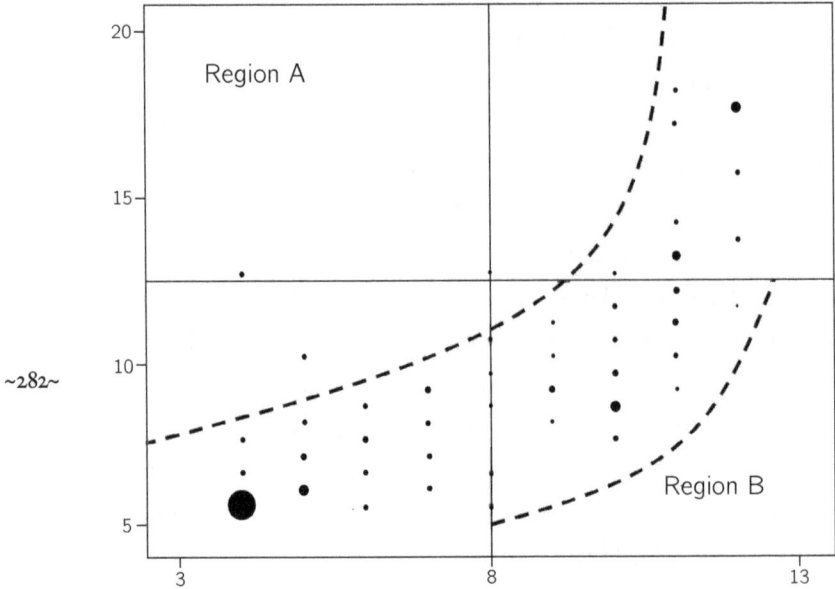

FIGURE 3 Empirical morphospace of cultural complexity for cases in the Atlas of Cultural Evolution (Peregrine 2003) dated from the last 12,000 years. The x-axis is the scale factor, and the y-axis is the technology factor (see Table 3); the size of the dots indicates the number of cases. The dashed lines demarcate two regions where empirical cases are extremely rare. The cases within region A are all pastoralists who obtained metals and other sophisticated technologies through interaction with larger sedentary societies.

be randomly distributed, but this is obviously not the case. High-scale, low-technology societies are essentially absent (region B), as are very low-scale, high-technology societies (region A). It would appear that technology and scale place constraints on one another in such a way that a particular scale is required to support particular technological capabilities or vice versa. Figure 4 demonstrates these constraints diachronically. Both scale and technology evolve together, filling in the morphospace in a roughly linear pattern that avoids both high-scale, low-technology and low-scale, high-technology realms. There appears to be interdependence between scale and technology such that neither can grow without the other growing in roughly parallel fashion. The evaluation of developmental and functional constraints should provide a way

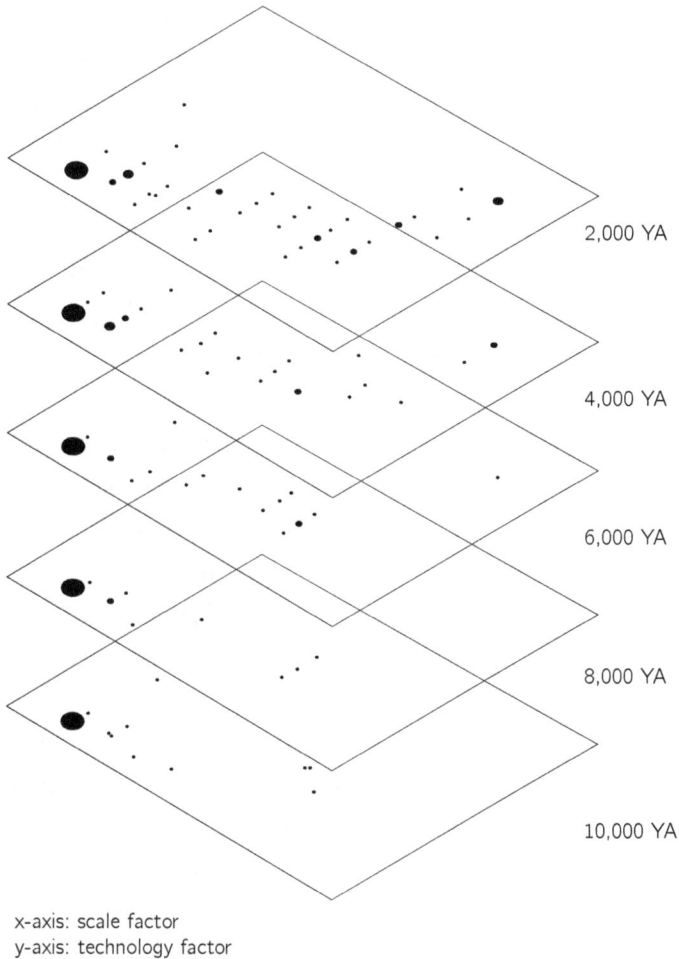

x-axis: scale factor
y-axis: technology factor

FIGURE 4 Empirical morphospace of societal scale and technology in 2,000-year intervals from 12,000 to 2,000 years ago. Size of dots indicates the number of cases.

of understanding both the empty regions and the relationship between scale and technology.

Functional constraints refer to forms that cannot exist because they are "lethal" to individuals of the given taxon—these forms simply do not function successfully (McGhee 2007:111). The primary functional constraint for region A may be the need for a sufficient population to construct, operate, and maintain more complex and diverse technologies; that is, with a small population, there are not enough people to share the time and expertise needed to maintain a complex technological regime. For region B, it seems likely that complex technologies are required to support a large population. Simple technologies cannot produce sufficient food, clothing, housing, and other resources for a large population.

Developmental constraints refer to forms within a given taxon that cannot exist because they lack the features necessary to produce specific forms (McGhee 2007:114). Following the work of Brian Arthur (2009), I hypothesize that with a small population there may not be sufficient "minds" to develop new technologies. According to Arthur (2009), technological innovations primarily grow through the combination of existing "modules." If there are not enough interacting minds with knowledge of these modules to provide opportunities for combination, then new technology will not develop. Technological evolution can fully emerge only when the size of the interacting minds is large enough. In the case of region B, it seems well established that technology is cumulative, and that retrogression toward simpler technologies is rare in human societies (e.g., Tomasello 2011). Thus, as more complex technology develops, we would expect it to continue to evolve.

The discussion of developmental and functional constraints provides a useful path toward understanding the empty regions in the Figure 3 morphospace, but it does little to aid our understanding of the regions where cultures appear clustered. Chapter 7

in this volume, by Ortman, Blair, and myself, provides some additional ideas about how scalar efficiencies might produce regularities in social formations. But, as I noted earlier, these regions may
represent the effect of convergent evolution. McGhee (2007:93–
96) suggests that the exploration of adaptive landscapes offers an
excellent method for examining this process, and I evaluate this
suggestion in the next section.

ADAPTIVE LANDSCAPES

Wright (1932) developed the idea of adaptive landscapes as part of   ~285~
his seminal approach to understanding natural selection through
the lens of genetic fitness. An adaptive landscape, like that presented in Figure 5, can be interpreted as representing the relative
fitness of particular genetic variations. Peaks are regions with high
fitness, and valleys are regions of low fitness. It is assumed that
organisms evolve through moving from regions of low fitness to
those of high fitness, in essence, climbing to peaks in the adaptive
landscape. In Wright's model, convergent evolution is expected

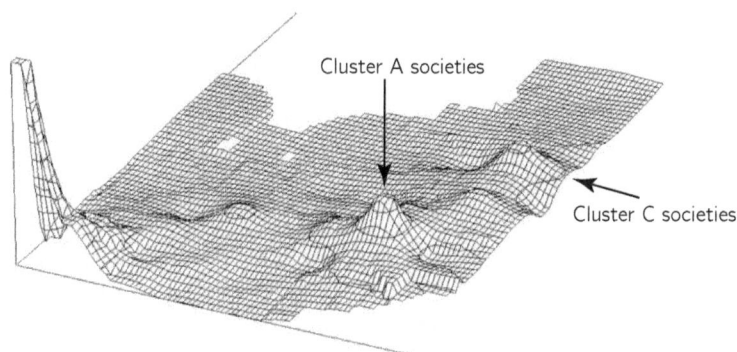

FIGURE 5 The empirical morphospace of Figure 3 displayed as a contour
map. This map can be interpreted as an adaptive landscape.

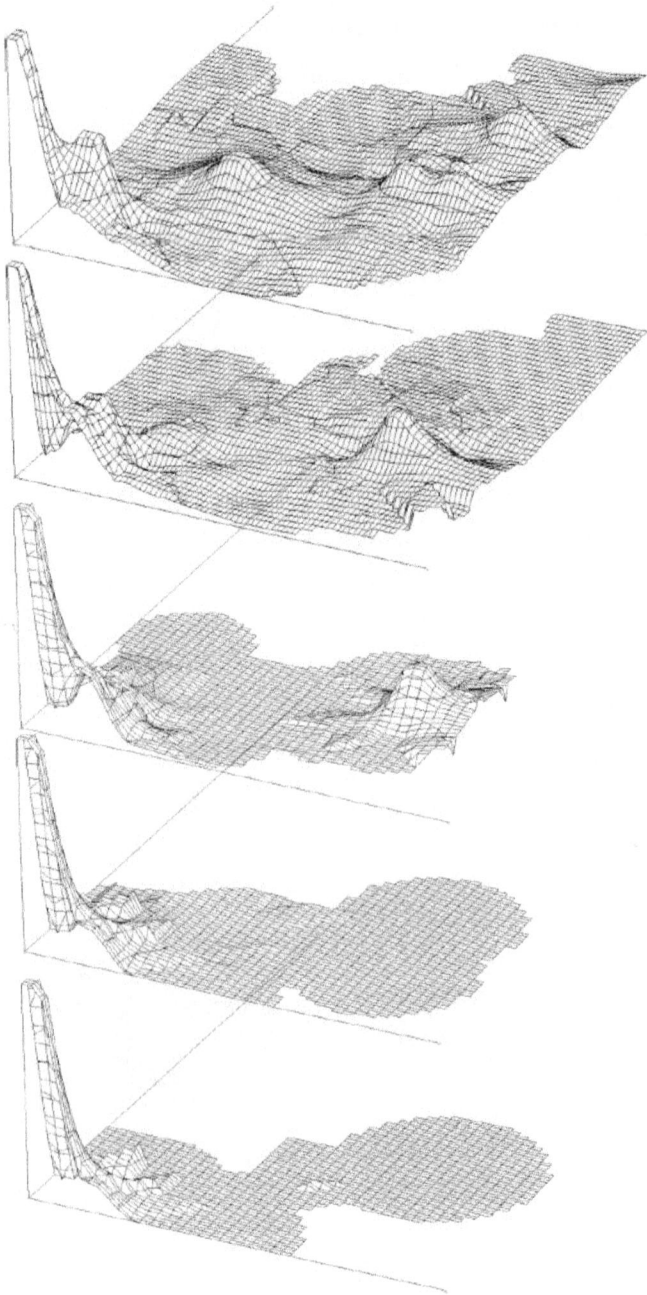

FIGURE 6 The empirical morphospaces of Figure 4 displayed as contour maps and interpreted as representing adaptive landscapes.

to occur regularly as different organisms move toward the same peak (McGhee 2007:33–36).

Convergence is thought to occur because there are limited ways of adapting to any particular environment, so similar species in a given environment will all eventually converge (McGhee 2011). Humans, however, shape our own environments, particularly after the adoption of agriculture (Smith 2007). We are niche constructors and environmental engineers; the environments we live in are (at least partially) our own creations (Laland and O'Brien 2011). Cultural innovations that we use to engineer the environment can lead to new opportunities and, in the context of adaptive landscapes, new adaptive peaks (e.g., Erwin 2008).

~287~

Indeed, humans have been called "the ultimate niche constructors" (Odling-Smee et al. 2003:28), as a primary focus of

The history of anthropology reflects an often-contentious dynamic between the study of unique cultural features and ones that are broadly shared across cultures.

cultural adaptation has been, at least since the emergence of agriculture, the active manipulation or engineering of the environment to create more stable or suitable conditions for agriculture (Smith 2007:197). As we create these new environments, we in turn provide a new environmental context that is inherited by descendants and by neighboring societies (Odling-Smee et al. 2003:2–16, 27, 252). Thus human niche construction produces not simply adaptive peaks but rather adaptive *attractors* that draw other societies toward them. This process of niche construction, ecological inheritance, and adaptive attraction is, I suggest, the source of recurrent social formations.

Figure 5 is a transformation of the empirical morphospace presented in Figure 3 into a three-dimensional adaptive landscape. Three adaptive peaks are clearly visible. The first, and largest, is at the origin of the landscape, where societies are small scale and have low technology. This represents the basic hunting-and-gathering adaptation upon which humans evolved and which remained a basic way of life for most humans until quite recently. A second peak reflects the societies of cluster A—sedentary, agricultural, and inegalitarian but non-state. These reflect the creation of a new adaptive peak based on agriculture.

Looking at Figure 6, which presents the adaptive landscapes diachronically, we can see the emergence of the cluster A adaptive peak and the convergence of many societies upon it. The peak first appears between 10,000 and 12,000 years ago. Between 8,000 and 10,000 years ago, there is divergence among newly agricultural societies, creating ripples in the adaptive landscape. Between 6,000 and 8,000 years ago, a new peak rapidly emerges as societies converge on the recurrent social formation identified here as cluster A.

Between 2,000 and 4,000 years ago, however, the cluster A adaptive peak itself begins to diversify, and a more rugged adaptive landscape appears in its environs. If this were occurring among a group of animal species, one would suggest that the development of greater morphological variety led to disruptive selection and/or that the new variations have created conflicting morphological or physiological constraints (McGhee 2007:15–21). I suggest that this is precisely what occurred with the development of agriculture. As agricultural lifestyles spread into new environments, and societies faced the emerging challenges of sedentary life, variations with conflicting constraints appeared, reflected here in a rugged adaptive landscape.

One of those new variations was the new peak that emerged between 2,000 and 4,000 years ago—a new peak representing the simple states of cluster C. Looking closely at the adaptive landscape for this period, one notices a long ridge connecting the cluster A and cluster C peaks. This ridge represents societies of increasing scale, and I suggest that it captures the evolutionary movement of societies converging on the new cluster C peak. If we were able to move the adaptive landscape forward another 1,000 years (the data prevent us from doing so), we should see the cluster C peak continuing to grow and the cluster A peak continuing to decline. Were we to look at an adaptive landscape of societies today, we would find a "Fujiyama" landscape with one large peak encompassing   ~289~ virtually all societies and focused on what is only a small hill at the very top right of the 2,000-to-4,000-years-ago landscape—the complex states of cluster D.

Toward a Theory of Recurrent Social Formations

The history of anthropology reflects an often-contentious dynamic between the study of unique cultural features and ones that are broadly shared across cultures (Carneiro 2003). The latter study has been marred by a history of ethnocentric perspectives and, more significantly, by a taxonomic approach lacking a general theory of cultural evolution. I have proposed one means to remedy problems in the taxonomic approach by applying methods used in evolutionary biology, thus linking the study of cultural variation to similar studies of phylogenetic variation (see also Currie 2013; Lipo et al. 2006; O'Brien et al. 2008). The methods of evolutionary biology provide a path toward a taxonomic approach to recurrent social formations that is both theoretically based and empirically robust (see also Mesoudi et al. 2006; O'Brien and Lyman 2000).

I am certainly not the first to suggest that the methods of evolutionary biology should play a larger role in archaeological research, but I do suggest that the work presented here is distinct in demonstrating, both *empirically* and *systematically*, the importance of comparative archaeology to understanding cultural evolution. Specifically, I have shown that large-scale comparative analyses of the archaeological record can be analyzed in the same way as large-scale comparative data on biological organisms. This comparative method is relatively rare in archaeology, where case studies are far more common, and almost nonexistent in the diachronic form employed here. Beyond that simple methodological insight, I have demonstrated *empirically* the presence of recurrent social formations that appear very similar to those put forward by anthropologists more than half a century ago. It is important to note that these recurrent social formations emerged from the analyses, not from a priori coding criteria or theory. They are, quite simply, present in the data (Sabloff and Cragg demonstrate a similar phenomenon in chapter 4 in this volume). The question remains, what processes create these recurrent social formations?

I have suggested that recurrent social formations are the result of convergent evolution in an adaptive landscape with several major peaks. These peaks reflect stable adaptations within particular physical and, more importantly, cultural environments (see also Laland and O'Brien 2010:308). But in understanding the appearance and nature of these peaks, it is essential to recognize that humans are the quintessential environmental engineers. Cultural innovations (such as agriculture) are niche constructors, effectively shaping the adaptive landscape. These innovations also create conflicting constraints, leading to diversity in specific adaptive forms. Thus, recurrent social formations are not identical but reflect the unique historical trajectory of societies as they are attracted toward specific adaptive peaks.

In closing, I return to the issue with which I opened this chapter—the issue of cultural taxonomy. Anthropology has long recognized that there are social formations that seem to recur, but there has been considerable controversy over how to identify and define, or even whether to identify and define, these recurrent social formations. I have illustrated an approach to this issue that appears to have identified inductively the presence of at least three recurrent social formations that have evolved within the last 10,000 years. This simple finding provides empirical evidence that the two most widely used cultural taxonomies—those of Fried (1967) and Service (1962)—represent, at least in a broad context, social reality. Thus I argue that work employing "traditional" cultural taxonomies—including work done by the Santa Fe Institute exploring the origins of states—has used appropriate concepts and should not be criticized out of hand for using those "traditional" taxonomic units. ❧

ACKNOWLEDGMENTS

This paper was born during a discussion with Doug Erwin about morphospace analysis and grew through the influence of many conversations in the stimulating environment of the Santa Fe Institute. I want to thank the many colleagues at the Santa Fe Institute who pointed me in fruitful directions and helped me to refine my ideas. I also want to thank George McGhee for his useful input, particularly on the concept of convergent evolution. Doug Erwin and George McGhee read an early version of this paper, and for that I again thank them. Parts of this paper were presented at the 2014 meetings of the American Anthropological Association and the Society for Anthropological Sciences, as well in colloquia at the Santa Fe Institute. This work was supported by a grant from the John Templeton Foundation to the Santa Fe Institute.

REFERENCES CITED

Arthur, W. Brian
2009   *The Nature of Technology: What It Is and How It Evolves.*
Free Press, New York.

Atkinson, Quentin, and Harvey Whitehouse
2011   The Cultural Morphospace of Ritual Form: Examining
Modes of Religiosity Cross-Culturally. *Evolution and Human
Behavior* 32:50–62.

Bailey, Kenneth
1994   *Typologies and Taxonomies: An Introduction to Classification
Techniques.* Sage, Thousand Oaks, California.

Blanton, Richard, Gary Feinman, Stephen Kowlewski, and Peter N.
Peregrine
1996   A Dual-Processual Theory for the Evolution of
Mesoamerican Civilization. *Current Anthropology* 37:1–14.

Bourdieu, Pierre
1977   *Outline of a Theory of Practice.* Trans. Richard Nice.
Cambridge University Press, Cambridge.

Carneiro, Robert
1962   Scale Analysis as an Instrument for the Study of Cultural
Evolution. *Southwestern Journal of Anthropology* 18:149–169.

1970   Scale Analysis, Evolutionary Sequences, and the Rating of
Cultures. In *A Handbook of Method in Cultural Anthropology*, edited
by Raoul Naroll and Ronald Cohen, pp. 834–871. Natural History
Press, Garden City, New York.

2003   *Evolutionism in Cultural Anthropology: A Critical History.*
Westview Press, Boulder, Colorado.

Chamberlain, John
1976   Flow Patterns and Drag Coefficients of Cephalopod Shells.
Paleontology 19:539–563.

Chick, Gary
1997   Cultural Complexity: The Concept and Its Measurement.
Cross-Cultural Research 31:275–307.

Currie, Thomas
2013   Cultural Evolution Branches Out: The Phylogenetic
Approach in Cross-Cultural Research. *Cross-Cultural Research*
47:102–130.

Erwin, Douglas
2008   Macroevolution of Ecosystem Engineering, Niche
Construction and Diversity. *Trends in Ecology and Evolution*
23:304–310.

Feinman, Gary, and Jill Neitzel
1984   Too Many Types: An Overview of Sedentary Prestate
Societies in the Americas. *Advances in Archaeological Method and
Theory* 7:39–102.

Foote, Michael
1990   Nearest-Neighbor Analysis of Trilobite Morphospace.
Systematic Zoology 39:371–382.

Fried, Morton   ~293~
1967   *The Evolution of Political Society: An Essay in Political
Anthropology*. Random House, New York.

Gell-Mann, Murray
2011   Regularities in Human Affairs. *Cliodynamics* 2:52–70.

Guttman, Louis
1950   The Basis for Scalogram Analysis. In *Measurement and
Prediction*, edited by Samuel A. Stouffer, pp. 60–90. Princeton
University Press, Princeton, New Jersey.

Laland, Kevin, and Michael O'Brien
2010   Niche Construction Theory and Archaeology. *Journal of
Archaeological Method and Theory* 17:303–322.

2011   Cultural Niche Construction: An Introduction. *Biological
Theory* 6:191–202.

Lipo, Carl, Michael O'Brien, and Stephen Shennan (editors)
2006   *Mapping Our Ancestors: Phylogenetic Methods in
Anthropology and Prehistory*. Aldine, New Brunswick, New Jersey.

McGhee, George
2007   *The Geometry of Evolution: Adaptive Landscapes and
Theoretical Morphospaces*. Cambridge University Press, Cambridge.

2011   *Convergent Evolution: Limited Forms Most Beautiful*. MIT
Press, Cambridge, Massachusetts.

McIver, John, and Edward Carmines
1981   *Unidimensional Scaling*. Sage, Thousand Oaks, California.

Mesoudi, Alex, Andrew Whiten, and Kevin Laland
2006   Towards a Unified Science of Cultural Evolution. *Behavioral
and Brain Sciences* 29:329–347.

Milner, George
1991   *The Cahokia Chiefdom*. Smithsonian, Washington, DC.

Morgan, Lewis Henry
1877   *Ancient Society.* Henry Holt, New York.

Murdock, George, and Catherine Provost
1973   Measurement of Cultural Complexity. *Ethnology* 12:379–392.

O'Brien, Michael, and R. Lee Lyman
2000   *Applying Evolutionary Archaeology: A Systematic Approach.* Springer, New York.

O'Brien, Michael, R. L. Lyman, M. Collard, C. J. Holden, R. D. Gray, and S. J. Shennan
2008   Phylogenetics and the Evolution of Cultural Diversity. In *Cultural Transmission and Archaeology*, edited by M. J. O'Brien, pp. 39–58. Society for American Archaeology, Washington, DC.

Odling-Smee, John, Kevin Laland, and Marcus Feldman
2003   *Niche Construction: The Neglected Process in Evolution.* Princeton University Press, Princeton, New Jersey.

Ortman, Scott, Andrew Cabaniss, Jennie Sturm, and Luis Bettencourt
2014   The Pre-History of Urban Scaling. *PloS One* 9(2):e87902. DOI:10.1371/journal.pone.0087902

Peebles, Christopher, and Susan Kus
1977   Some Archaeological Correlates of Ranked Societies. *American Antiquity* 42:421–448.

Peregrine, Peter
2001   *Outline of Archaeological Traditions.* HRAF Press, New Haven, Connecticut.

2003   Atlas of Cultural Evolution. *World Cultures* 14(1):2–88.

Peregrine, Peter, Carol Ember, and Melvin Ember
2004   Universal Patterns in Cultural Evolution: An Empirical Analysis Using Guttman Scaling. *American Anthropologist* 106(1):145–149.

Peregrine, Peter, and Melvin Ember (editors)
2001–2002   *Encyclopedia of Prehistory.* 9 vols. Kluwer Academic/ Plenum Publishers, New York.

Raup, Donald
1966   Geometric Analysis of Shell Coiling: General Problems. *Proceedings of the National Academy of Sciences* 47:602–609.

Sahlins, Marshall
1960   Evolution: Specific and General. In *Evolution and Culture*, edited by Marshall Sahlins and Elman Service, pp. 12–44. University of Michigan Press, Ann Arbor.

Service, Elman
1962   *Primitive Social Organization: An Evolutionary Perspective.*
Random House, New York.

Smith, Bruce
2007   Niche Construction and the Behavioral Context of Plant
and Animal Domestication. *Evolutionary Anthropology* 16:188–199.

Steward, Julian
1955   *Theory of Culture Change: The Methodology of Multilinear
Evolution.* University of Illinois Press, Urbana.

Tomasello, Michael
2011   Human Culture in Evolutionary Perspective. *Advances in
Culture and Psychology* 1:5–51.   ~295~

Tylor, Edward Burnett
1871   *Primitive Culture: Researches into the Development of
Mythology, Philosophy, Religion, Art, and Custom.* John Murray,
London.

Wright, Sewall
1932   The Roles of Mutation, Inbreeding, Crossbreeding and
Selection in Evolution. *Proceedings of the Sixth International Congress
of Genetics* 1:356–366.

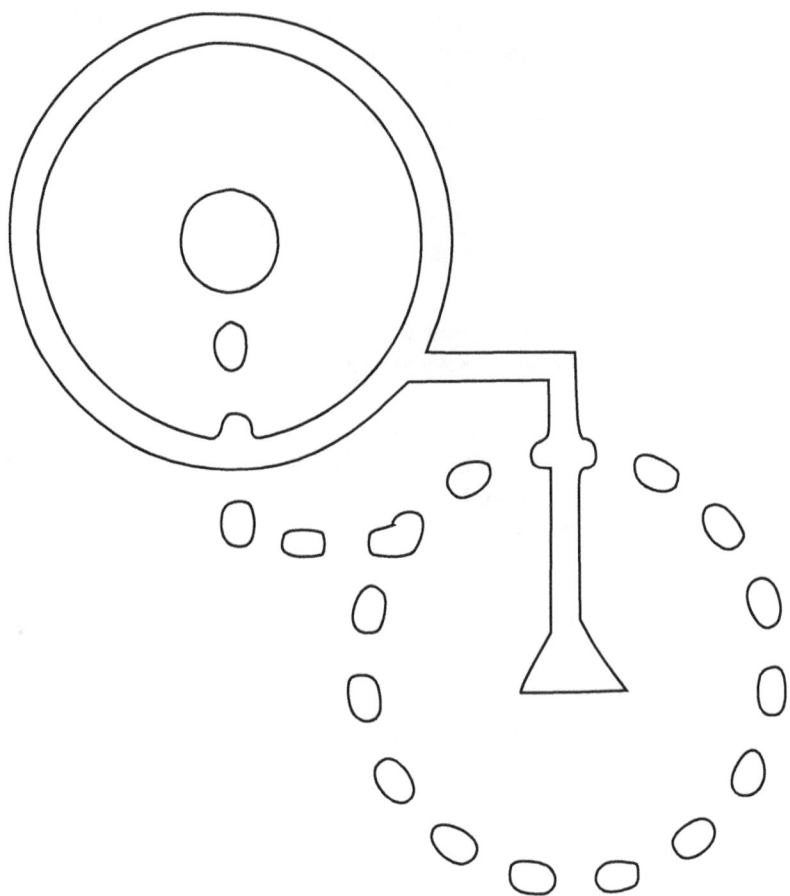

oj いう

CONCLUDING REMARKS

Paula L.W. Sabloff, Santa Fe Institute
and Jeremy A. Sabloff, Santa Fe Institute and University of Pennsylvania

> The most important issue confronting the social sci-
> ences is the extent to which human behaviour is shaped
> by factors that operate cross-culturally as opposed to
> factors that are unique to particular cultures.
>
> —BRUCE G. TRIGGER (2003)

~297~

Archaeologists and historians have been interested in explaining the rise of premodern states since the nineteenth century. Not surprisingly, a plethora of hypotheses have been advanced in attempts to rigorously explain the development of state society through time and space. But despite all the efforts of archaeological field-workers and theoreticians alike, no widely accepted theory has yet emerged. Even the best of the large-scale efforts at examining the causes of the rise of states, such as Bruce Trigger's monumental *Understanding Early Civilizations* (2003), have not been able to create widely accepted theories (also see Johnson and Earle 2000).

Given this intellectual situation, we have organized this volume in the hope of advancing our understanding of the emergence of premodern states by using the methods and techniques of complexity science—network analysis, agent-based modeling, and the like—in new ways. Part I (chapters 2–3) provides historical context, discussing past efforts at formulating theories on the emergence of premodern states. The authors define key terms and review some of the problems that proponents of earlier theories faced. Part II (chapters 4–7) presents some of the complexity-science approaches that are used to uncover new patterns in the extant archaeological data. And Part III (chapters 8–9) offers some initial syntheses of data patterns. These three sections enhance old approaches to theory building by combining complexity-science strategies with theories and methods used by archaeologists in the recent past.

What approaches have scholars used to advance our understanding of the archaeological record of premodern states? First, they have been conducting both intensive and extensive surveys and excavations of a particular site or tradition. They have used their findings, as well as other archaeological studies of the tradition, to analyze and interpret the data, sometimes employing a wide variety of foci, from environment to demography to subsistence to warfare to trade. For example, in their pathbreaking Valley of Mexico survey, Sanders et al. (1979) focused particularly on the environment, demography, and subsistence in their examination of growing complexity in the region. Further south, Flannery and Marcus (1983) looked at aspects of all these factors in their analysis of the evolution of the Zapotec state.

Second, scholars have been applying theories developed in other disciplines such as political science, political philosophy, evolutionary biology, and sociology/anthropology to see if these theories add to their understanding of the archaeological record. For instance, scholars have widely used the writings of Hobbes, Marx, Foucault, Bourdieu, and Giddens to help them understand the evolution of complex society (Dobres and Robb 2000; Dornan 2002; Roes and Raymond 2003; Robb 2010). Another recent example is Blanton and Fargher's application of collective action theory from the field of political science to state formation (Blanton and Fargher 2008, 2016; also see Carballo and Feinman 2016). They ask how and why collective states (societies composed of rulers and taxpayers) arise, and where and why they differ in organization (Blanton and Fargher 2008:14–15). They then compare 30 societies to answer their questions. They found "generally strong statistical support for collective action theory" (2008:251) even though they noted problems in applying the model to historical states (2008:254).

Third, comparative analyses of particular emerging or full-blown states, as well as the collapse of states, have begun to yield a greater understanding of cross-cultural regularities. Multistate comparisons such as Tainter (1988), Trigger (2003), Schwartz and Nichols (2006), Sandweis and Quilter (2008), Blanton and Fargher (2008), McAnany and Yoffee (2010), Jennings (2016), and Johnson (2017), among many others, have been undertaken, as have two- or three-state comparisons that may or may not have focused on particular issues. Adams (1967) was a modern pioneer in this regard, comparing Mesopotamia and Central Mexico. Marcus (2006) compares New Kingdom Egypt, the Late Classic Maya Lowlands, and Postclassic Zapotec to see if elites used the same strategies. Some examples of the comparative approach are Gillespie and Joyce (1997), who contrast Indonesia with the Maya; Blanton et al.'s (1996) intra-Mesoamerican state comparisons; and chapters in Feinman and Marcus's edited volume, *Archaic States* (1998), that compare two or more early states, such as Baines and Yoffee's comparison of Egypt and Mesopotamia.

Fourth, archaeologists have begun to use tools developed or used in complexity science, such as network analysis, agent-based modeling, computer simulations, and evolutionary game theory, to find such patterns in new ways (see Sabloff 1981; Spencer 1998; Kohler and Gumerman 2000; Axtell et al. 2002; Kohler and Van der Leeuw 2007; Rogers and Cegielski 2017). The chapters in this volume, especially those in Part II, show the efficacy of using complexity tools for pattern recognition in archaeology. We hope that the approaches discussed in those chapters will stimulate more archaeologists to use these tools for comparative analysis in order to isolate cross-cultural patterns. Pattern recognition, or the uncovering of regularities in different unconnected cases, is an essential early step in the formulation and testing of hypotheses about the processes of growing complexity in premodern state formation.

While the use of research tools from complexity science will not necessarily provide new theories for the rise of early complex societies, they can be productively used to generate new working hypotheses and to test alternate hypotheses. Pattern recognition is the baseline; when combined with theoretical understanding from earlier archaeological studies and other disciplines, the meaning of the patterns for the emergence of premodern states may be better understood.

Despite such efforts, the resulting hypotheses have yet to receive widespread acceptance. In light of these concerns, the contributors to this volume have taken a step back by examining the possible utility of new approaches in moving the theory-building effort forward in new, productive ways. Our ultimate goal is the same as earlier authors—to explain the development of premodern states—but, pragmatically, we believe that more modest intermediate goals are appropriate at this time. We hope this volume will stimulate further theory-building undertakings.

Last, it is worth asking why it is worth spending time (and taxpayer money!) learning how and why premodern states emerged from chiefdoms. Why did it take so long for states to arise? Why did they emerge independently on different continents and exhibit similar patterns of sociopolitical and economic organization? And once the state emerged more than five millennia ago, why has this form of organization lasted until the present day? Obviously, one answer is simply intellectual curiosity about the past and the nature of complex societies. But an even more compelling reason, we firmly believe, is that new understandings of the processes that led to the development of early states can provide historical contexts for key challenges that modern states around the world face today. For example, archaeologists now have better understanding of how elites gain and maintain economic and political control over populations. Our insights can enlighten discussions of the origins of today's deep divide between the rich (the 1 percent) and everyone else (the 99 percent): how did societies come to this point, no matter what their specific political organization? Our data can enrich people's understanding of the key role that organized religion and religious beliefs have played in bolstering states and their leaders. It can also shed important light on the successes and failures of different uses of environments over the long terms. In short, the study of premodern states can certainly be relevant to today's problems. It is not just an arcane pursuit! ❧

~301~

REFERENCES CITED

Adams, Robert McC.
1967   *Evolution of Urban Society.* Aldine, Chicago.

Axtell, Robert L., Joshua M. Epstein, Jeffrey S. Dean, George J. Gumerman, Alan C. Swedlund, Jason Harburger, Shubha Chakravarty, Ross Hammond, Jon Parker, and Miles Parker
2002   Population Growth and Collapse in a Multiagent Model of the Kayenta Anasazi in Long House Valley. *Proceedings of the National Academy of Sciences* 99:7275–7279.

Baines, John, and Yoffee, Norman
2008   Order, Legitimacy, and Wealth in Ancient Egypt and Mesopotamia. In *Archaic States,* edited by Gary Feinman and Joyce Marcus, pp. 199–260. School of American Research, Santa Fe, New Mexico.

Blanton, Richard E., Gary M. Feinman, Stephen A. Kowalewski, and Peter N. Peregrine
1996   A Dual-Processual Theory for the Evolution of Mesoamerican Civilization. *Current Anthropology* 17(1):1–85.

Blanton, Richard L., and Lane Fargher
2008   *Collective Action in the Formation of Pre-Modern States.* Springer, New York.

Blanton, Richard E., with Lane F. Fargher
2016   *How Humans Cooperate: Confronting the Challenges of Collective Action*. University Press of Colorado, Denver.

Carballo, David M., and Gary M. Feinman
2016   Cooperation, Collective Action, and the Archeology of Large-Scale Societies. *Evolutionary Anthropology* 25:288–296.

Dobres, Marcia-Anne, and John Robb
2000   Agency in Archaeology: Paradigm or Platitude? In *Agency in Archaeology*, edited by Marcia-Anne Dobres, and John Robb, pp. 3–17. Routledge, London.

Dornan, J. L.
2002   Agency and Archaeology: Past, Present, and Future Directions. *Journal of Archaeological Method and Theory* 9:303–329.

Feinman, Gary, and Joyce Marcus (editors)
2008   *Archaic States*. School of American Research, Santa Fe, New Mexico.

Flannery, Kent, and Joyce Marcus (editors)
1983   *The Cloud People*. Academic Press, New York.

Gillespie, Susan D., and Rosemary A. Joyce
1997   Gendered Goods: The Symbolism of Maya Hierarchical Exchange Relations. In *Women in Prehistory: North America and Mesoamerica*, edited by Cheryl Claassen and Rosemary A. Joyce, pp. 189–207. University of Pennsylvania Press, Philadelphia.

Jennings, Justin
2016   *Killing Civilization: A Reassessment of Early Urbanism and Its Consequences*. University of New Mexico Press, Albuquerque.

Johnson, Allen W., and Timothy Earle
2000   *The Evolution of Complex Societies*. Stanford University Press, Stanford, California.

Johnson, Scott A. J.
2017   *Why Did Ancient Civilizations Fail?* Routledge, New York.

Kohler, Timothy A., and George G. Gumerman (editors)
2000   *Dynamics in Human and Primate Societies: Agent-Based Modeling of Social and Spatial Processes*. Santa Fe Institute, Santa Fe, New Mexico.

Kohler, Timothy A., and Sander E. van der Leeuw
2007   *The Model-Based Archaeology of Socionatural Systems*. School for Advanced Research, Santa Fe, New Mexico.

Marcus, Joyce
2006   Identifying Elites and Their Strategies. In *Intermediate Elites in Precolumbian States and Empires*, edited by C. M. Elson and R. A. Covey, pp. 212–246. University of Arizona Press, Tucson.

Robb, John
2010   Beyond Agency. *World Archaeology* 42:493–520.

Roes, Frans, and Michael Raymond
2003   Belief in Moralizing Gods. *Evolution and Human Behavior* 24:126–135.

Rogers, J. Daniel, and Wendy H. Cegielski
2017   Building a Better Past with the Help of Agent-based Modeling. *Proceedings of the National Academy of Sciences* 49:12841–12844.

Sabloff, Jeremy A. (editor)
1981   *Simulations in Archaeology*. School of American Research, Santa Fe, New Mexico.

Sanders, William T., Jeffrey R. Parsons, and Robert S. Santley
1979   *The Basin of Mexico: Ecological Processes in the Evolution of a Civilization*. Academic Press, New York.

Sandweis, Daniel H., and Jeffrey Quilter (editors)
2008   *El Niño: Catastrophism, and Culture Change in Ancient America*. Dumbarton Oaks, Washington, DC.

Schwartz, Glenn M., and John J. Nichols (editors)
2006   *After Collapse: The Regeneration of Complex Societies*. University of Arizona Press, Tucson.

Spencer, Charles S.
1998   A Mathematical Model of Primary State Formation. *Cultural Dynamics* 10:5–20.

Tainter, Joseph A.
1988   *The Collapse of Complex Societies*. Cambridge University Press, Cambridge.

Trigger, Bruce G.
2003   *Understanding Early Civilizations*. Cambridge University Press, Cambridge.

APPENDIX

APPENDIX TO CHAPTER 5 Predicted effects of socioecology on strategy frequencies, alliance size, and inequality given low, medium, and high patchiness.[1]

Landscape patchiness var(μ_k)	Parameter		Strategy frequencies						Mean alliance size	Inequality Gini(W)
			Non-territorial β	Territorial β	Alliance-forming β	Cooperative β	Coordinated punisher β	Hierarchical β	β	β
Low	(Intercept)	–	1.35 ***	-1.35 ***	-0.06	0.43 ***	0.97 ***	-0.22 *	-0.57 ***	0.10 ***
	Decisiveness of alliances	$\log(\alpha)$	-0.02	0.02	0.05	0.03	0.13	0.03	-0.06	0.01
	Cost of ally	c^A	-0.07 .	0.07 .	0.02	-0.01	0.10	0.02	-0.05	0.01
	Cost of cooperation	c^C	-0.10 *	0.10 *	0.05	0.10	0.13	0.04	-0.05	0.01
	Cost of enforcement	c^E	-0.05	0.05	0.04	-0.02	0.06	0.01	-0.03	0.01
	Effic. of political process	d	0.06	-0.06	-0.07	-0.13	-0.33 *	-0.03	0.09	0.00

[1] Notes: Models estimating strategy frequencies and alliance sizes were fit using the lmer function in the lme4 package in R. Models estimating inequality were fit using the lm function. Continuous predictor and response variables (except for inequality) were normalized to have mean 0 and standard deviation 1. The reported standardized regression coefficients (β's) thus indicate a change of β standard deviations in the response variable with an increase of 1 standard deviation in the predictor variable. $n = 20{,}000$ generations in 200 runs for each model (with random effects for run), except for inequality, where $n = 200$ runs (without random effects). *** $p < 0.001$; ** $p < 0.01$; * $p < 0.05$; . $p < 0.1$.

APPENDIX TO CHAPTER 5 (continued)

Landscape patchiness var(μ_k)	Parameter		Strategy frequencies							
			Non-territorial β	Territorial β	Alliance-forming β	Cooperative β	Coordinated punisher β	Hierarchical β	Mean alliance size β	Inequality Gini(W) β
Medium	(Intercept)	—	-0.27 ***	0.27 ***	-0.15	-0.43 ***	-0.20	-0.02	0.10	0.23 ***
	Decisiveness of alliances	log(α)	-0.01	0.01	0.16 .	0.03	0.08	0.23 *	0.20 *	0.01
	Cost of ally	c^A	-0.03	0.03	-0.73 ***	-0.04	-0.22 .	-0.27 *	-0.54 ***	-0.02
	Cost of cooperation	c^C	-0.04	0.04	-0.40 ***	-0.09	0.03	-0.69 ***	-0.49 ***	-0.01
	Cost of enforcement	c^E	-0.04	0.04	0.06	0.01	-0.04	0.04	0.03	0.01
	Effic. of political process	d	-0.03	0.03	-0.07	0.02	-0.02	0.05	0.00	0.01
High	(Intercept)	—	-0.72 ***	0.72 ***	0.26 ***	-0.36 ***	-0.16 .	0.32 ***	0.44 ***	0.33 ***
	Decisiveness of alliances	log(α)	0.01	-0.01	0.18 **	0.07	0.01	0.28 ***	0.13 .	0.06 ***
	Cost of ally	c^A	-0.10 **	0.10 **	-0.82 ***	-0.04	-0.24 **	-0.45 ***	-0.75 ***	-0.03 *
	Cost of cooperation	c^C	-0.06 *	0.06 *	-0.29 ***	-0.16 **	-0.06	-0.69 ***	-0.21 **	-0.05 ***
	Cost of enforcement	c^E	-0.04	0.04	-0.13 .	0.00	-0.01	-0.13 .	-0.16 *	-0.05 ***
	Effic. of political process	d	-0.02	0.02	0.00	0.01	0.02	0.13 .	0.02	0.01

APPENDIX TO CHAPTER 7 Data for Archaeological Traditions (refer to page 191). Pop = largest settlement population; Area = largest settlement area (ha).

Number	Name	Sequence	Start (BP)	End (BP)	AgDate (BP)	Pop	Area (ha)
1001	Early Paleo-Indian		12200	11000		30	1
1005	Late Paleo-Indian		11000	6000		30	1
1010	Paleo-Arctic		11000	6000		30	1
1013	Late Tundra		8000	6000		100	1
1015	Northern Archaic		6000	4000		30	1
1020	Western Arctic Small Tool		4700	2500		30	1
1025	Norton		3000	1000		200	1
1030	Thule		2100	100		300	1
1035	Eastern Arctic Small Tool		4000	2700		20	1
1040	Dorset		2800	700		75	1
1045	Shield Archaic		6000	3000		30	1
1050	Initial Shield Woodland		3000	600		35	1
1055	Terminal Shield Woodland		600	150			
1065	Northwest Microblade		7000	2000		25	1
1070	Proto-Athapaskan		2000	150		50	1
1075	Eastern North America Early Archaic		10000	8000		30	1
1080	Eastern North America Middle Archaic		8000	6000		50	1
1085	Eastern North America Late Archaic		6000	3000		50	1
1115	Eastern North America Early Woodland		3000	2100	2100	150	1
1120	Adena		3000	2100	2100	50	1

Data for Archaeological Traditions (*continued*)

ACE Score	NPP Mean	NPP Variance	NPP Min	NPP Max	QALY	15P5	Diversity
10	0.381	0.077	0.006	1.321			3
10	0.348	0.039	0.001	1.321			3
10	0.226	0.003	0.020	0.443			1
10	0.334	0.003	0.086	0.450			2
10	0.307	0.008	0.019	0.504			1
10	0.280	0.005	0.086	0.450			2
12	0.280	0.005	0.086	0.450			14
10	0.278	0.005	0.086	0.450			14
10	0.167	0.000	0.113	0.190			4
11	0.167	0.000	0.113	0.190			4
10	0.315	0.009	0.042	0.568			4
13	0.335	0.012	0.042	0.617			7
13							
10	0.336	0.009	0.019	0.504			1
13	0.315	0.011	0.019	0.504			7
10	0.655	0.004	0.535	0.931			5
10	0.558	0.015	0.089	0.931		0.217	5
12	0.607	0.010	0.089	0.931		0.254	11
14	0.599	0.010	0.089	0.931	20.55	0.194	12
16	0.700	0.001	0.631	0.756			15

Data for Archaeological Traditions (*continued*)

Number	Name	Sequence	Start (BP)	End (BP)	AgDate (BP)	Pop	Area (ha)
1125	Eastern North America Middle Woodland		2100	1300	2100	100	160
1130	Hopewell		2100	1700	2100	50	48
1140	Northeast Middle Woodland		2400	1000	1000	200	10
1155	Eastern North America Late Woodland		1300	500	1000	344	2.6
1160	Northeast Late Woodland	Iroquois	1000	500	1000	1950	6
1175	Mississippian	Eastern US	1100	500	2100	12500	830
1180	Oneota	Eastern US	1000	230	2100	300	500
1185	Fort Ancient	Eastern US	1000	200	2100	500	20
1195	Proto-Iroquois	Iroquois	950	350	1000	1000	3
1205	Plains Archaic		9000	2500		30	1
1210	Plains Woodland		2500	200		50	1
1215	Central Plains Village		1050	150	1050	100	1
1216	Northern Plains Village		950	150		2120	16
1220	Early Desert Archaic		10000	8000		25	1
1225	Middle Desert Archaic	Hohokam	8000	2000	2000	67	6
1230	Fremont	Pueblo	1600	500	2300	400	42
1235	Late Desert Archaic		1600	150		30	1
1240	San Dieguito		10000	8000		30	1
1260	High Plains Early and Middle Archaic		8000	3000		30	1
1270	High Plains Late Archaic		3000	1500		50	1
1275	High Plains Late Prehistoric		1500	150		96	
1300	Early Hohokam	Hohokam	2000	900	2000	500	50

ACE Score	NPP Mean	NPP Variance	NPP Min	NPP Max	QALY	15P5	Diversity
16	0.673	0.004	0.535	0.931	22.37	0.344	21
16	0.698	0.001	0.600	0.789			23
14	0.603	0.002	0.426	0.687		0.171	14
15	0.667	0.003	0.543	0.781	20.24	0.284	29
18	0.603	0.002	0.426	0.687	20.40	0.280	14
23	0.679	0.003	0.535	0.931		0.285	42
19	0.577	0.007	0.334	0.702			26
19	0.697	0.000	0.669	0.731	19.22	0.322	26
21							26
10	0.373	0.032	0.001	0.732			2
14	0.384	0.029	0.001	0.698			5
15	0.359	0.019	0.034	0.684	20.21	0.289	14
15	0.292	0.029	0.009	0.586			19
10	0.047	0.002	0.001	0.205			4
12	0.120	0.018	0.000	0.659		0.136	7
18	0.055	0.002	0.006	0.191		0.297	5
11	0.072	0.002	0.012	0.205			2
10	0.064	0.004	0.000	0.221			1
10	0.203	0.026	0.002	0.493			2
11	0.195	0.024	0.002	0.493			6
12	0.179	0.021	0.002	0.448			
19	0.012	0.000	0.000	0.044		0.305	20

Data for Archaeological Traditions (*continued*)

Number	Name	Sequence	Start (BP)	End (BP)	AgDate (BP)	Pop	Area (ha)
1305	Late Hohokam	Hohokam	900	500	2000	632	106
1310	Early Mogollon	Mogollon	2000	1000	2000	300	3
1315	Late Mogollon	Mogollon	1000	600	2000	800	9
1320	Basketmaker	Pueblo	2300	1300	2300	100	6
1325	Early Anasazi	Pueblo	1300	700	2300	804	38
1330	Late Anasazi	Pueblo	700	450	2300	2833	15
1335	Patayan	Hohokam	1600	500	2000	800	1
1350	Windmiller		8000	3000		30	1
1355	Cosumnes		3000	1500		25	1
1360	Hotchkiss		1500	150		300	1
1365	Early Southern California		8000	3000		300	1
1370	Late Southern California		3000	150		800	3
1385	Early Sierra Nevada		10000	3000		30	1
1395	Late Sierra Nevada		3000	150		60	1
1400	Cascade		8000	5000		30	1
1405	Tucaunon		5000	2500		60	1
1410	Harder		2500	500		150	1
1430	Archaic Oregon Coast		10000	2000		50	1
1435	Formative Oregon Coast		2000	150		300	1
1440	Early Northwest Coast		9500	5500		30	1
1445	Middle Northwest Coast		5500	1500		500	1
1450	Late Northwest Coast		1500	200		2000	1
1460	Aleutian		6000	250		330	4
1463	Ocean Bay		8000	4000		200	1
1465	Kodiak		4000	700		200	1
1475	Trincheras	Trincheras	3000	450	3000	2100	10

~312~

ACE Score	NPP Mean	NPP Variance	NPP Min	NPP Max	QALY	15P5	Diversity
20						0.233	21
19	0.120	0.004	0.010	0.281		0.251	15
19	0.120	0.004	0.010	0.281		0.341	17
14	0.089	0.009	0.001	0.282		0.272	6
18	0.083	0.008	0.001	0.282	15.76	0.222	26
19					14.62	0.293	25
14	0.069	0.003	0.012	0.214			6
10	0.294	0.010	0.180	0.463			4
11	0.294	0.010	0.180	0.463			4
12	0.294	0.010	0.180	0.463			8
13	0.268	0.016	0.064	0.486	20.98	0.045	11
18	0.268	0.016	0.064	0.486	20.07	0.065	17
10	0.079	0.003	0.019	0.162			1
18	0.080	0.002	0.019	0.162			12
11	0.305	0.029	0.002	0.547			6
12	0.305	0.029	0.002	0.547			2
14	0.298	0.029	0.002	0.547			2
10	0.524	0.003	0.432	0.589			4
18	0.524	0.003	0.432	0.589			15
11	0.466	0.008	0.295	0.737			2
15	0.453	0.008	0.295	0.737			17
20	0.453	0.008	0.295	0.737			22
12							11
11	0.362	0.002	0.271	0.414			2
11	0.362	0.002	0.271	0.414			12
20	0.343	0.035	0.007	0.659			17

Data for Archaeological Traditions (*continued*)

Number	Name	Sequence	Start (BP)	End (BP)	AgDate (BP)	Pop	Area (ha)
1480	Huatabampo	Huatabampo	1800	500	1800	150	3
1485	Coahuila		8000	500		30	1
1505	Early Mesoamerican Archaic		9600	7000		25	1
1510	Highland Mesoamerican Archaic	Central Mexico	7000	4000	4000	30	1
1512	Lowland Mesoamerican Archaic	Maya	7000	3800	3800	50	2
1515	Highland Mesoamerican Early Preclassic	Central Mexico	4000	2600	4000	530	50
1520	Highland Mesoamerican Late Preclassic	Central Mexico	2600	1700	4000	20000	450
1522	Preclassic Maya	Maya	3800	1850	3800	30000	100
1525	Olmec	Central Mexico	3400	2100	4000	18000	200
1530	West Mexico Classic	Trincheras	1800	1100	3000	8500	101
1535	Southern Mexican Highlands Classic	Central Mexico	1700	1300	4000	17242	442
1540	Central Mexico Classic	Central Mexico	2100	1300	4000	95597	225
1545	Gulf Coast Classic	Central Mexico	2100	800	4000	30000	500
1550	Classic Maya	Maya	1850	1100	3800	55000	350
1560	Postclassic Maya	Maya	1100	400	3800	12000	400
1565	West Mexico Postclassic	Trincheras	1100	480	3000	20000	500
1570	Central Mexico Postclassic	Central Mexico	700	430	4000	212500	135

ACE Score	NPP Mean	NPP Variance	NPP Min	NPP Max	QALY	15P5	Diversity
18	0.293	0.031	0.007	0.593			11
10	0.217	0.013	0.001	0.643			1
10	0.692	0.038	0.360	1.097			2
11	0.582	0.020	0.360	0.929			7
11	0.797	0.033	0.374	1.097			8
19	0.584	0.021	0.360	0.929	18.04	0.177	33
20	0.542	0.013	0.360	0.790	17.44	0.093	35
22	0.797	0.033	0.374	1.097			38
19	0.518	0.001	0.482	0.554			26
23							35
23	0.494	0.008	0.362	0.594			43
23	0.666	0.008	0.501	0.757	16.28	0.121	40
23	0.715	0.032	0.482	0.929			36
25	0.789	0.034	0.374	1.097	15.99	0.186	46
25	0.797	0.033	0.374	1.097	17.87	0.091	47
23	0.529	0.010	0.344	0.697			46
28	0.686	0.007	0.501	0.790	17.42	0.225	51

Data for Archaeological Traditions (*continued*)

Number	Name	Sequence	Start (BP)	End (BP)	AgDate (BP)	Pop	Area (ha)
1575	Southern Mexican Highlands Postclassic	Central Mexico	1300	430	4000	9250	200
2005	Old South American Hunting-Collecting		13000	9000			
2010	Old Amazonian Collecting-Hunting		11000	7000			
2015	Late Andean Hunting-Collecting		9000	7000			
2030	Early Northwest South American Littoral		9000	5500	3500		
2035	Late Northwest South American Littoral		5500	3000			
2040	Early Caribbean	Caribbean	3000	1000	3000		
2045	Late Caribbean	Caribbean	1000	500	3000	3000	
2050	Early Amazonian		7000	2000			
2055	Late Amazonian	Amazonian	2000	50	2000		
2060	Early Parana-Pampean		7000	1500			
2065	Late Parana-Pampean	Parana	1500	500	1500	300	
2070	Magellan-Fuegan		6300	50			
2075	Early East Brazilian Uplands		11000	5000			
2080	Late East Brazilian Uplands		5000	50		150	1.2
2085	Sambaqui		7000	500			
2100	Tupi	Tupi	1500	150	1500	6000	
2105	Early Chibcha	Chibcha	3500	1200	3500	900	

ACE Score	NPP Mean	NPP Variance	NPP Min	NPP Max	QALY	15P5	Diversity
23	0.519	0.009	0.362	0.618			50
10	0.393	0.122	0.006	1.081			
10	0.866	0.013	0.454	1.204			
10	0.383	0.120	0.006	1.081			
11	0.843	0.080	0.000	1.321	21.20	0.241	
11	0.817	0.080	0.000	1.321			
18	0.736	0.041	0.360	1.089			
19	0.715	0.050	0.360	1.089			
10	0.871	0.011	0.417	1.204			
18	0.869	0.011	0.417	1.204			
10	0.421	0.046	0.003	0.936			
19	0.417	0.047	0.003	0.936			
10	0.501	0.000	0.498	0.504			
10	0.623	0.053	0.006	1.146			
14	0.597	0.045	0.006	1.095	23.60	0.048	
12	0.632	0.091	0.101	1.146			
18	0.643	0.067	0.006	1.146			
18	0.892	0.077	0.000	1.321	20.17	0.178	

Data for Archaeological Traditions (*continued*)

Number	Name	Sequence	Start (BP)	End (BP)	AgDate (BP)	Pop	Area (ha)
2110	Late Chibcha	Chibcha	1200	500	3500	5000	
2115	Ecuadorian Coast Regional Development	Chibcha	2500	1500	3500		
2120	Manteno	Chibcha	1200	410	3500		
2125	Ecuadorian Highlands	Ecuador	1500	500	1500		
2145	Cocle	Chibcha	2000	400	3500	2000	80
2150	Chiriqui	Chibcha	1200	500	3500	600	
2155	Paya	Chibcha	1500	500	3500		
2160	Nicoya	Nicoya	3600	500	3600		65
2200	Coastal Andean Archaic		7000	4100	4100		
2205	Highland Andean Early Archaic	Inca	7000	4500	3500	50	
2215	Highland Andean Late Archaic	Inca	4500	3500	3500	200	
2220	Coastal Andean Early Formative	Coastal Andean	4100	3000	4100	2000	111
2222	Coastal Andean Late Formative	Coastal Andean	3000	2200	4100	3000	200
2225	Highland Andean Formative	Inca	3500	2200	3500	250	
2230	Chavin	Inca	2800	2200	3500	2500	42
2235	Andean Regional Development	Inca	2200	1300	3500		
2240	Moche	Coastal Andean	1950	1200	4100	7500	13
2245	Nazca	Coastal Andean	2200	1200	4100		25
2250	Tiahuanaco	Inca	1600	900	3500	17500	60
2255	Huari	Inca	1200	900	3500	20000	60

ACE Score	NPP Mean	NPP Variance	NPP Min	NPP Max	QALY	15P5	Diversity
22	0.927	0.057	0.343	1.260	18.74	0.144	
22							
22	0.587	0.064	0.208	0.862			
20	1.020	0.000	1.020	1.020			
19							
19	1.072	0.008	0.982	1.162			
18	0.868	0.026	0.557	1.048			
21	0.621	0.000	0.621	0.621			
11	0.217	0.086	0.006	1.027	18.09	0.169	
11	0.404	0.116	0.006	1.081			
14	0.415	0.117	0.006	1.081			
21	0.217	0.086	0.006	1.027			
21							
21	0.415	0.117	0.006	1.081			
21							
24	0.494	0.129	0.004	1.081		0.234	
24							
24							
24	0.306	0.090	0.004	0.957		0.116	
24	0.677	0.144	0.000	1.321			

Data for Archaeological Traditions (*continued*)

Number	Name	Sequence	Start (BP)	End (BP)	AgDate (BP)	Pop	Area (ha)
2260	Andean Regional States	Inca	900	530	3500	10000	24
2265	Chimu	Inca	1050	480	3500	27500	2000
2270	Aymara Kingdoms	Inca	900	530	3500		
2275	Inca	Inca	800	468	3500	100000	
2280	South Andean Ceramic	Inca	2500	500	3500		
3005	Aurignacian		40000	25000		25	
3010	Perigordian		30000	22000		100	
3020	Solutrean		22000	18000			
3025	Magdalenian		18000	11000			
3030	Eastern European Mesolithic		11000	6000	6000		
3040	Western European Mesolithic		11000	6000	6000		
3045	Southeastern European Mesolithic	SE Europe	11000	8000	8000		
3050	Southeastern European Neolithic	SE Europe	8000	6500	8000	300	
3060	Impressed Ware	Europe	6800	6000	6800	200	
3065	Linear Pottery	Europe	6500	6000	6800	200	7
3075	European Megalithic	Europe	6000	4500	6800	1000	5
3080	Corded Ware	Europe	6000	3800	6800		
3085	Bell Beaker	Europe	4500	3650	6800	1250	
3090	European Early Bronze Age	Europe	4700	3800	6800	100	
3095	Western European Earlier Bronze Age	Europe	3800	3300	6800		
3100	Western European Later Bronze Age	Europe	3300	2800	6800	300	

ACE Score	NPP Mean	NPP Variance	NPP Min	NPP Max	QALY	15P5	Diversity
24					19.10	0.308	
24							
24							
25							
20	0.257	0.079	0.006	1.027	17.00	0.226	
10							
10	0.376	0.040	0.000	0.777			
10							
10	0.624	0.013	0.107	0.911			
11	0.403	0.028	0.000	0.777		0.189	
11	0.632	0.014	0.107	0.911		0.205	
11	0.561	0.005	0.432	0.700		0.164	
19	0.559	0.004	0.432	0.700		0.232	
19						0.323	
19							
18	0.610	0.021	0.107	0.911		0.270	
19	0.477	0.015	0.207	0.761		0.327	
19	0.610	0.021	0.107	0.911		0.217	
20	0.638	0.006	0.405	0.819			
21							
21	0.628	0.015	0.107	0.911			

Data for Archaeological Traditions (*continued*)

Number	Name	Sequence	Start (BP)	End (BP)	AgDate (BP)	Pop	Area (ha)
3105	West-Central European Early Iron Age	Europe	2800	2400	6800	5000	100
3110	West-Central European Late Iron Age	Europe	2400	2033	6800	50	
3115	Northeastern European Bronze Age	Europe	3500	2800	6800		
3120	Northeastern European Iron Age	Europe	2800	2000	6800		
3125	East-Central European Iron Age	Europe	2800	2000	6800		
3130	Southeastern European Early Chalcolithic	SE Europe	6500	5500	8000	150	
3131	Southeastern European Late Chalcolithic	SE Europe	5500	4500	8000	12000	345
3135	Southeastern European Bronze Age	SE Europe	5100	3100	8000	6400	32
3140	Scandinavian Neolithic		6000	3800			
3145	Scandinavian Bronze Age	Scandinavian	3800	2500	3800		
3150	Roman Iron Age	Europe	2033	1500	6800	200	1.5
3155	Scandinavian Iron Age	Scandinavian	2500	1500	3800		
3165	Romano-British	Europe	2100	1500	6800	28000	134
3170	Caucasian Neolithic	Caucasian	8000	6500	8000	2500	5
3175	Caucasian Chalcolithic	Caucasian	6500	5500	8000	200	5
3180	Caucasian Bronze Age	Caucasian	5600	2700	8000		100
4005	Ordosian	Shang	40000	8500	8500		
4015	Peiligang	Shang	8500	6200	8500		2

ACE Score	NPP Mean	NPP Variance	NPP Min	NPP Max	QALY	15P5	Diversity
24							
18							
22	0.492	0.015	0.207	0.690			
24	0.508	0.012	0.257	0.690			
24	0.608	0.006	0.408	0.731			
21	0.601	0.008	0.425	0.819			
22							
28	0.548	0.004	0.432	0.700			
16	0.417	0.009	0.257	0.659			
24	0.420	0.008	0.257	0.659			
21	0.682	0.001	0.615	0.742			
29	0.421	0.008	0.257	0.659			
29	0.624	0.004	0.475	0.685			
19	0.507	0.016	0.137	0.777			
21	0.506	0.017	0.137	0.777			
21	0.520	0.014	0.137	0.777			
10	0.651	0.058	0.003	1.098			
18	0.626	0.022	0.003	1.049			

Data for Archaeological Traditions (*continued*)

Number	Name	Sequence	Start (BP)	End (BP)	AgDate (BP)	Pop	Area (ha)
4020	Southeast China Early Neolithic		9000	5500		50	
4025	Yangshao	Shang	7000	4500	8500	300	
4030	Dawenkou	Shang	6200	4500	8500		
4035	Hongshan	Hongshan	7000	4500	7000	600	
4040	Daxi	Daxi	7000	4500	7000		
4045	Majiabang	Majiabang	7000	5000	7000		3
4055	Longshan	Shang	4500	3900	8500	20000	90
4060	Shang	Shang	3900	3100	8500	100000	325
4070	Early Xiajiadian	Hongshan	4500	3600	7000	2000	8.5
4075	Late Xiajiadian	Hongshan	3600	2500	7000		
4080	Southeast China Late Neolithic	SE China	5500	2500	5500		
4105	Chulumn	Chulumn	8000	4000	8000	100	
4110	Mumum	Chulumn	4000	2300	8000	201	
4150	Japanese Upper Paleolithic		20000	12000			
4155	Jomon		12000	2500		500	
4160	Yayoi	Yayoi	2500	1500	2500		
4300	Siberian Early Upper Paleolithic		42000	28000		50	
4301	Siberian Middle Upper Paleolithic		28000	19000		100	
4302	Siberian Late Upper Paleolithic		19000	10000			
4305	Amur Paleolithic		30000	12000			
4310	Amur Neolithic and Bronze		12000	1500		240	
4320	Baikal Neolithic and Bronze		8000	3000			

ACE Score	NPP Mean	NPP Variance	NPP Min	NPP Max	QALY	15P5	Diversity
15	0.859	0.017	0.307	1.098			
19	0.535	0.049	0.013	1.049			
19	0.671	0.014	0.359	0.777			
19	0.470	0.018	0.181	0.687			
19	0.853	0.047	0.332	1.088			
19	0.898	0.020	0.713	1.098			
25	0.542	0.041	0.013	1.049			
28							
20	0.480	0.016	0.181	0.687			
20	0.472	0.017	0.181	0.687			
20	0.865	0.031	0.307	1.098			
18	0.730	0.002	0.669	0.803			
22	0.455	0.014	0.004	0.803			
10							
12	0.753	0.005	0.677	1.080			
26	0.753	0.005	0.677	1.080			
10							
10							
10	0.292	0.012	0.001	0.496			
11							
12	0.461	0.005	0.322	0.597			
12	0.358	0.009	0.015	0.496			

Data for Archaeological Traditions (*continued*)

Number	Name	Sequence	Start (BP)	End (BP)	AgDate (BP)	Pop	Area (ha)
4340	Siberian Neolithic and Bronze		10000	2100		75	
4350	Siberian Protohistoric		2100	500		50	
4360	Holocene Stone Age of Northeast Asia		10500	3000			
4370	Kamchatka Mesolithic		7000	4000			
4375	Tarya Neolithic		4000	2500			
4380	Old Itel'man		2500	500			
4400	Kelteminar		8000	4000		200	
4405	Eurasian Steppe Nomad		6500	4000			
4410	Andronovo	Andronovo	4000	2800	4000	4000	
4420	Early Nomad		2800	2300			
4425	Late Nomad		2300	1500			
4430	Scythian-Sarmatian		4000	1700			
4500	Eastern Central Asia Paleolithic		40000	6000			
4510	Eastern Central Asia Neolithic and Bronze		6000	1500			1
5005	Lapita		3500	2000		750	
5010	Fijian	Polynesian	2100	200	2100		
5015	Samoan	Polynesian	2000	200	2100		
5020	Tongan	Polynesian	2000	200	2100		
5030	Marquesan	Polynesian	1700	175	2100		
5035	Hawaiian	Polynesian	800	200	2100	900	25
5040	Tahitian	Polynesian	1400	200	2100		
5045	Easter Island	Polynesian	1500	400	2100	400	
5050	Maori	Polynesian	900	200	2100	2500	
5070	New Guinea Neolithic	New Guinea	10000	100	10000	100	
5075	Melanesian		2500	200		200	

ACE Score	NPP Mean	NPP Variance	NPP Min	NPP Max	QALY	15P5	Diversity
12	0.300	0.008	0.002	0.570			
13	0.281	0.005	0.017	0.570			
12	0.273	0.003	0.020	0.392			
13	0.305	0.001	0.252	0.372			
13	0.305	0.001	0.252	0.372			
13	0.305	0.001	0.252	0.372			
14	0.165	0.020	0.001	0.488			
13	0.386	0.026	0.002	0.678			
19	0.138	0.019	0.002	0.487			
17							
17	0.140	0.020	0.002	0.488			
17	0.329	0.032	0.000	0.777			
10	0.191	0.047	0.003	1.117			
14	0.192	0.048	0.003	1.117			
16							
22							
22							
22							
22							
22							
22							
22							
22							
19	0.905	0.142	0.000	1.264			
18							

Data for Archaeological Traditions (*continued*)

Number	Name	Sequence	Start (BP)	End (BP)	AgDate (BP)	Pop	Area (ha)
5080	Micronesian		3000	200			
5085	Early Australian		50000	7000			
5090	Late Australian		7000	200			
5100	Southeast Asia Upper Paleolithic		40000	10000		30	
5105	Hoabinhian		10000	4000			
5110	Southeast Asia Neolithic and Bronze	SE Asia	6500	2500	6500		
5115	Island Southeast Asia Late Prehistoric	SE Asia	2500	1500	6500	900	5
5120	Mainland Southeast Asia Late Prehistoric	SE Asia	2500	1500	6500		500
5505	South Asian Upper Paleolithic	Indus	30000	7000	7000		
5510	South Asian Microlithic		7000	3500			
5515	Indus Neolithic	Indus	7000	5000	7000		
5520	Ganges Neolithic	Ganges	4000	2500	4000	200	
5525	South Indian Neolithic	South India	5000	3100	5000		
5530	Early Indus	Indus	5000	4600	7000		27.3
5535	Mature Indus	Indus	4600	3900	7000	60000	150
5540	Central Indian Neolithic	Central India	5000	3100	5000		
5550	Gangetic India	Ganges	2500	2000	4000		
5555	Vedic	Indus	3900	2000	7000		
5560	Central Indian Iron Age	Central India	3100	2100	5000		
5565	South Indian Iron Age	South India	3100	2100	5000		
6005	Aterian		100000	8000			

ACE Score	NPP Mean	NPP Variance	NPP Min	NPP Max	QALY	15P5	Diversity
18							
10	0.208	0.028	0.001	0.904			
12	0.208	0.028	0.001	0.904			
10							
10	0.770	0.029	0.307	1.144			
20	0.730	0.029	0.323	1.144		0.279	
22	1.041	0.004	0.927	1.178			
26	0.721	0.020	0.307	1.122		0.228	
10	0.243	0.043	0.001	0.949			
12	0.274	0.042	0.001	0.949			
19	0.126	0.024	0.000	0.526			
18	0.539	0.033	0.292	0.989			
18	0.130	0.008	0.031	0.317			
24	0.132	0.026	0.000	0.526			
30	0.132	0.026	0.000	0.526			
21	0.300	0.046	0.001	0.949			
30	0.507	0.022	0.292	0.944			
30	0.132	0.026	0.000	0.526			
22	0.168	0.011	0.001	0.359			
22	0.294	0.046	0.001	0.913			
10	0.145	0.027	0.014	0.815			

Data for Archaeological Traditions (*continued*)

Number	Name	Sequence	Start (BP)	End (BP)	AgDate (BP)	Pop	Area (ha)
6010	Late Pleistocene– Early Holocene Maghreb		20000	7500			
6025	Southern Mediterranean Neolithic		7500	4000			
6030	Neolithic of Capsian		7500	4000			
6035	North African Protohistoric		4000	3000			
6040	Saharo- Sudanese Neolithic		8000	3000			
6060	Southern and Eastern Africa LSA		40000	2000		30	
6065	Wilton		10000	2000			
6075	South African Iron Age	S Africa	2000	500	2000	14500	720
6080	East African Microlithic		20000	5000		30	
6090	East African Neolithic	E Africa	5000	1200	5000	50	
6095	Early Khartoum		10000	5700			
6100	Khartoum Neolithic	Khartoum	5700	3550	5700	300	4
6110	Late Paleolithic Egypt	Egypt	45000	7000	7000		
6115	Upper Egypt Predynastic	Egypt	7000	5000	7000	10000	30
6125	Lower Egypt Predynastic	Egypt	7000	5000	7000	2000	100
6130	Early Dynastic Egypt	Egypt	5000	4700	7000	30000	110
6135	Protohistoric Egypt	Egypt	4700	3552	7000	50000	350
6160	Tshitolian		10000	4000		30	
6165	Central African Neolithic	Central Africa	4000	2000	4000		2

ACE Score	NPP Mean	NPP Variance	NPP Min	NPP Max	QALY	15P5	Diversity
10	0.108	0.018	0.000	0.506			
14	0.164	0.023	0.000	0.506			
14	0.054	0.009	0.013	0.285			
21	0.072	0.013	0.013	0.506			
15	0.167	0.028	0.014	0.837			
10	0.277	0.039	0.001	0.855			
10	0.280	0.046	0.009	0.846			
21	0.321	0.050	0.009	0.855			
10	0.327	0.082	0.001	1.309			
16	0.334	0.081	0.001	1.309			
10	0.006	0.000	0.004	0.107			
17	0.007	0.001	0.004	0.112			
10	0.016	0.003	0.001	0.282			
17	0.021	0.004	0.000	0.221			
17	0.002	0.000	0.001	0.024			
30	0.017	0.004	0.001	0.282			
30							
10	0.878	0.038	0.104	1.225			
17	0.897	0.036	0.104	1.225			

Data for Archaeological Traditions (*continued*)

Number	Name	Sequence	Start (BP)	End (BP)	AgDate (BP)	Pop	Area (ha)
6170	Central African Iron Age	Central Africa	2000	500	4000		4
6175	West African LSA	West Africa	40000	4000	4000		
6180	West African Neolithic	West Africa	4000	2000	4000		
6185	West African Iron Age	West Africa	2500	1200	4000		
6190	West African Regional Development	West Africa	1200	630	4000	50000	
6195	Nachikufan		16000	2000		25	
6205	Epipaleolithic	Mesopotamia	45000	10100	9000	50	
6210	Natufian	Mesopotamia	12500	10100	9000	200	
6215	Aceramic Neolithic	Mesopotamia	10500	7500	9000	2500	15
6220	Ceramic Neolithic	Mesopotamia	8000	6100	9000	5000	13.5
6225	Chalcolithic	Arabia	6100	5500	6100		20
6230	Early Bronze Age	Arabia	5500	4000	6100	3700	15
6235	Middle Bronze Age	Arabia	4000	3595	6100	20000	80
6245	Halafian	Mesopotamia	7500	7000	9000	900	18
6250	Ubaid	Mesopotamia	7500	6000	9000	2250	15
6255	Late Chalcolithic Mesopotamia	Mesopotamia	6000	5100	9000	10000	70
6260	Jemdet Nasr	Mesopotamia	5100	4900	9000	20000	250
6265	Early Dynastic Mesopotamia	Mesopotamia	4900	4334	9000	60000	550
6270	Akkadian	Mesopotamia	4334	4112	9000	60000	550
6300	Arabian Upper Paleolithic		40000	11000			
6305	Early Arabian Pastoral		11000	5750			
6310	Middle Arabian Pastoral		5750	4200			

ACE Score	NPP Mean	NPP Variance	NPP Min	NPP Max	QALY	15P5	Diversity
21	0.893	0.038	0.104	1.225			
10	0.649	0.056	0.013	1.309			
16	0.631	0.053	0.065	1.097			
20	0.314	0.090	0.014	1.309			
23	0.344	0.071	0.014	1.309			
10	0.563	0.020	0.229	0.855			
10	0.314	0.025	0.003	0.651		0.293	
13	0.262	0.026	0.056	0.569		0.264	
16	0.260	0.030	0.000	0.651		0.365	
17	0.244	0.030	0.000	0.651			
23	0.224	0.030	0.006	0.569			
26	0.337	0.025	0.006	0.651			
30	0.336	0.025	0.006	0.651			
21							
23	0.268	0.030	0.000	0.651			
28	0.267	0.030	0.000	0.651			
30	0.114	0.009	0.000	0.335			
30							
30							
10	0.034	0.004	0.004	0.254			
11	0.032	0.004	0.004	0.254			
12	0.032	0.004	0.004	0.254			

Data for Archaeological Traditions (*continued*)

Number	Name	Sequence	Start (BP)	End (BP)	AgDate (BP)	Pop	Area (ha)
6315	Late Arabian Pastoral		4200	3595			
6320	Early Arabian Littoral		11000	7000			
6325	Middle Arabian Littoral		7000	3300			
6330	Late Arabian Littoral	Arabia	3300	2300	6100		
6340	Southern Asia Upper Paleolithic		40000	12000			
6350	Iranian Mesolithic		20000	11500		50	
6355	Iranian Neolithic	Iran	11500	7500	11500	400	
6360	Iranian Chalcolithic	Iran	7000	5000	11500	4000	20
6365	Iranian Bronze Age	Iran	5500	3500	11500	20000	
6375	Iranian Iron Age	Iran	3500	2550	11500		
7005	Oldowan		2300000	1600000			
7010	Acheulean		1800000	200000			
7015	Zhoukoudianian		600000	200000			
7020	Southern and Eastern Africa MSA		200000	40000			
7025	Middle Paleolithic		200000	40000			
7030	East Asian Middle Paleolithic		200000	40000			
7040	Middle Paleolithic Egypt		200000	40000			
7050	Siberian Mousterian		130000	42000			

ACE Score	NPP Mean	NPP Variance	NPP Min	NPP Max	QALY	15P5	Diversity
12	0.032	0.004	0.004	0.254			
11	0.048	0.007	0.000	0.251			
17	0.045	0.006	0.000	0.251			
28	0.045	0.006	0.000	0.251			
10							
10	0.406	0.059	0.004	1.264			
20	0.393	0.079	0.004	1.180			
20	0.385	0.093	0.004	1.180			
26	0.418	0.140	0.004	1.180			
30	0.227	0.076	0.014	1.180			
10							
10							
10							
10							
10							
10							
10							
10							

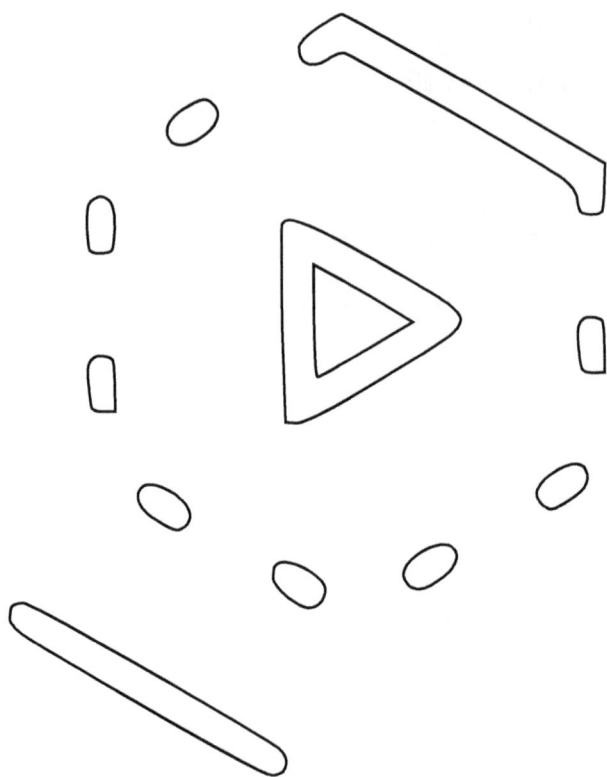

INDEX

A

accumulations research 244
active-passive dichotomy 94
Adams, Robert McC. 8, 18, 299
adaptation 77, 191
adaptive attraction 287, 290
adaptive attractors 287
adaptive landscape analysis 275
adaptive landscapes 273, 275, 285–289, 290
adaptive peaks 285–290, 290
administrators 68, 154; *See also* GOVERNMENT
Africa 39; *See also* SPECIFIC COUNTRIES AND REGIONS
agency
    concepts of 222
    of rulers 79
agent-based modeling 141, 297, 299
agents
    agent types 144
        merging and 168, 169
        warfare and 168, 169
    alliance-forming 112
    hierarchical 112
    nonhierarchical 112
    territorial 112
agriculture 26, 68, 70, 287, 290
    agricultural dependence 190, 191, 198–199
    agricultural societies 288
    agriculturists 105–106; *See also* FARMERS
    development of 288, 288–289
    environmental engineering and 287
    farmland 124, 148
    intensive 124
    irrigation agriculture 18, 24
alliance formation 78, 81, 106, 112–113, 120, 122, 124
    alliance-forming agents 112
    alliance-forming strategies 122

D

I

nonproducers 154; *See also* ADMINISTRATORS
non-states 63, 70, 77, 93, 94
non-state societies 280
nonterritorial agents 110
nonterritorial strategies 111
nonurban settlement systems 235
norms 237, 238, 239, 251
North America 191; *See also* SPECIFIC TRADITIONS
economies of scale in 202–203
list of attributes coded for 192–193
Northern Rio Grande region (New Mexico) 246
Northwest Coast 125, 174
novelty, emergence of 225; *See also* INNOVATION, INEVITABILITY OF
Nowak, M. 240
nuclear-family pattern 247

O

objects of study 21
obligations 175
Oceania 40
officials 68
offspring 115
Olson, Mancur 136
oral testimony 16–30
Ortman, Scott 9–8, 138–185, 285
Outline of Archaeological Traditions (OAT) 39, 40, 41, 42, 44, 274
working draft of 41
Outline of Cultural Materials 40
overrepresentation 37
owîngeh ("village") 248

P

Pacific peoples 19
paleopathology data 197–195
Panama 126
patches 110, 112, 114–115
patchiness 111, 120–121, 124, 306
path dependence 173
patronage 78, 81
pattern recognition 299–300
payoffs 144
the people, concept of 63
Peregrine, Peter 10, 41–49, 82, 189, 274–275, 276, 278, 285

U

UCINET
   core-periphery 81
   UCINET 6 62, 83, 87, 92
unit of analysis 44
universal laws, quest to establish 33
Upham, Steadman 139
urbanism 26
urbanization, degree of 190
urban scaling theory 233–236
urban society 21
urban systems 21
   diversity of professions in 203
   economies of scale and 200–202
   explaining emergence of 23
   explanations for emergence of 26
Uruk period 17, 23
US Southwest 40; *See also* SPECIFIC TRADITIONS AND REGIONS
   archaeological traditions of 190
   Mesa Verde region of 133–185
   pre-Hispanic 133–185
usufruct rights 70, 85
Utah 246

V

Valley of Mexico survey 298
values 238, 239, 249–250, 251; *See also* NORMS
Village Ecodynamics Project (VEP) I area 138–185
violence 19, 150; *See also* CONFLICT; WARFARE
   decline in 237–238, 246
voluntary participation 142, 143

W

Walker, Robert B. 19
Ward's method dendogram 280
Ware, John 61
warfare 18, 54, 150
   agent types and 168, 169
   as public good 142
   cooperation and 142
   high-ranking women and 72–73
   marriage alliances and 95
   over access to prime sites 106

ACKNOWLEDGMENTS

A number of colleagues helped with the project that made this volume possible. Henry Wright was an inspiration from day 1 of the project to its conclusion. In addition to the chapter authors and to those who are acknowledged in the individual chapters, we particularly wish to thank Eric Rupley, Anne Kandler, Karl LaFavre, Hannah DeRose-Wilson, and Charles Perreault for all their assistance. Our SFI colleagues David Krakauer, Jessica Flack, Geoffrey West, Luis Bettencourt, Sam Bowles, Chip Stanish, Doug Erwin, George Gumerman, the late Linda Cordell, and Sander van der Leeuw, among many others, offered much appreciated advice, as did Chuck Spencer, Colin Renfrew, Gary Feinman, and Linda Nicholas. Elisabeth Johnson, Ginny Greninger, Hilary Skolnik, Ginger Richardson, Laura Egley Taylor, and Lucy Fleming assisted the project in a number of ways. We also thank Rachel Fudge for her helpful copyediting and Heather Dubnick for the index.

This publication was made possible through the support of a grant from the John Templeton Foundation. We especially wish to thank Mary Ann Meyers, Barnaby Marsh, and all the program officers at the Foundation for their enthusiastic support and advice.

The SFI Press would not exist without the support of
William H. Miller and the Miller Omega Program,
Andrew Feldstein and the Feldstein Program on History,
Law, and Regulation, and Alana Levinson-Labrosse.

JEREMY A. SABLOFF, an archaeologist, is an External Professor and Past President of the Santa Fe Institute and the Christopher H. Browne Distinguished Professor of Anthropology, Emeritus at the University of Pennsylvania. He received his B.A. from the University of Pennsylvania (1964) and his PhD from Harvard University (1969). He is a member of the National Academy of Sciences and the American Philosophical Society and a Fellow of the American Academy of Arts and Sciences, the Society of Antiquaries, London, and the American Association for the Advancement of Science. His principal scholarly interests include ancient Maya civilization, the rise of complex societies and cities, the history of archaeology, and the relevance of archaeology in the modern world. Previous books include *Excavations at Seibal; Ceramics* (1975), *The Cities of Ancient Mexico* (1989; 2nd ed., 1997), *The New Archaeology and the Ancient Maya* (1990), and *Archaeology Matters* (2008).

PAULA L.W. SABLOFF, an anthropologist, is an External Professor at the Santa Fe Institute. She received her B.A. from Vassar College (1967) and her M.A. and PhD from Brandeis University (1971 and 1977 respectively). She has taught at the University of Pennsylvania and the University of Pittsburgh while conducting research in Mexico, the United States and Mongolia. A political anthropologist, her research interests have focused on Mexicans' manipulation of government land classification and Mongolians' changing ideas of democracy and capitalism. For the past five years, she has returned to her first love—archaeology—creating a database in order to compare premodern societies. She is writing papers from this database on kings' modes of risk reduction in war and the political agency of royal women. She has written *Does Everyone Want Democracy? Insights from Mongolia* (2013) and *Conversations with Lew Binford: Drafting the New Archaeology* (1998). She has edited *Mapping Mongolia: Situating Mongolia in the World from Geologic Time to the Present* (2011), *Modern Mongolia: Reclaiming Genghis Khan* (2001), *Higher Education in the Post-communist World: Eight Case Studies* (1999), and *Reform and Change in Higher Education: International Perspectives* (1995).

⅍ PR❧·SS

The Santa Fe Institute Press endeavors to communicate the best of complexity science and to capture a sense of the diversity, range, breadth, excitement, and ambition of SFI-related research. To provide a distillation of discussions, debates, and meetings across a range of influential and nascent topics. To change the way we think.

SEMINAR SERIES
New findings emerging from the Institute's ongoing working groups and research projects, for an audience of interdisciplinary scholars and practitioners.

ARCHIVE SERIES
Fresh editions of classic texts from the complexity canon, spanning the Institute's thirty years advancing the field.

COMPASS SERIES
Provoking, exploratory volumes aiming to build complexity literacy in the humanities, industry, and the curious public.

For forthcoming titles, inquiries, or news about the Press, contact us at sfipress@santafe.edu.

COLOPHON

The body copy for this book was set in EB Garamond, a typeface designed by Georg Duffner after the Ebenolff-Berner type specimen of 1592. Headings are in Kurier, a typeface created by Janusz M. Nowacki, based on typefaces by the Polish typographer Małgorzata Budyta. Additional type is set in Cochin, a typeface based on the engravings of Nicolas Cochin, for whom the typeface is named.

The SFI Press Complexity Glyphs used throughout this book were designed by Brian Crandall Williams.

SANTA FE INSTITUTE
COMPLEXITY
GLYPHS

ZERO

ONE

TWO

THREE

FOUR

FIVE

SIX

SEVEN

EIGHT

NINE

-A-

-B-

-C-

-D-

-E-

-F-

-G-

-H-

-I-

-J-

-K-

-L-

-M-

-N-

-O-

-P-

-Q-

-R-

-S-

-T-

-U-

-V-

-W-

-X-

-Y-

-Z-

SEMINAR SERIES

www.ingramcontent.com/pod-product-compliance
Lightning Source LLC
Chambersburg PA
CBHW031505180326
41458CB00044B/6699/J